HiSET Subject Test

Mathematics

Student Practice Workbook

+ Two Realistic HiSET Math Tests

Math Notion

www.MathNotion.com

HiSET Subject Test – Mathematics

HiSET Subject Test – Mathematics

HiSET Subject Test Mathematics

Published in the United State of America By

The Math Notion

Web: WWW.MathNotion.com

Email: info@Mathnotion.com

Copyright © 2021 by the Math Notion. All rights reserved. No part of this publication may be reproduced, stored in a retrieval system, or transmitted in any form or by any means, electronic, mechanical, photocopying, recording, scanning, or otherwise, except as permitted under Section 107 or 108 of the 1976 United States Copyright Ac, without permission of the author.

All inquiries should be addressed to the Math Notion.

ISBN: 978-1-63620-042-2

HiSET Subject Test – Mathematics

The Math Notion

Michael Smith has been a math instructor for over a decade now. He launched the Math Notion. Since 2006, we have devoted our time to both teaching and developing exceptional math learning materials. As a test prep company, we have worked with thousands of students. We have used the feedback of our students to develop a unique study program that can be used by students to drastically improve their math scores fast and effectively. We have more than a thousand Math learning books including:

– **GED Math Prep**

– **TABE Math Prep**

– **TASC Math Prep**

– **Accuplacer Math Prep**

– **Common Core Math Prep**

–**many Math Education Workbooks, Study Guides, Practice and Exercise Books**

As an experienced Math test preparation company, we have helped many students raise their standardized test scores—and attend the colleges of their dreams: We tutor online and in person, we teach students in large groups, and we provide training materials and textbooks through our website and through Amazon.

You can contact us via email at:

info@Mathnotion.com

HiSET Subject Test – Mathematics

Get the Targeted Practice You Need to Ace the HiSET Math Test!

HiSET Subject Test - Mathematics includes easy-to-follow instructions, helpful examples, and plenty of math practice problems to assist students to master each concept, brush up their problem-solving skills, and create confidence.

The HiSET math practice book provides numerous opportunities to evaluate basic skills along with abundant remediation and intervention activities. It is a skill that permits you to quickly master intricate information and produce better leads in less time.

Students can boost their test-taking skills by taking the book's two practice HiSET Math exams. All test questions answered and explained in detail.

Important Features of the HiSET Math Book:

- A **complete review** of HiSET math test topics,
- Over 2,500 practice problems covering all topics tested,
- The most important concepts you need to know,
- Clear and concise, easy-to-follow sections,
- Well designed for enhanced learning and interest,
- Hands-on experience with all question types
- **2 full-length practice tests** with detailed answer explanations
- Cost-Effective Pricing

Powerful math exercises to help you avoid traps and pacing yourself to beat the HiSET test. Students will gain valuable experience and raise their confidence by taking math practice tests, learning about test structure, and gaining a deeper understanding of what is tested on the HiSET Math. If ever there was a book to respond to the pressure to increase students' test scores, this is it.

WWW.MathNotion.COM

… So Much More Online!

- ✓ FREE Math Lessons
- ✓ More Math Learning Books!
- ✓ Mathematics Worksheets
- ✓ Online Math Tutors

For a PDF Version of This Book

Please Visit WWW.MathNotion.com

HiSET Subject Test – Mathematics

Contents

Chapter 1 : Integers and Number Theory 11
   Rounding 12
   Whole Number Addition and Subtraction 13
   Whole Number Multiplication and Division 14
   Rounding and Estimates 15
   Adding and Subtracting Integers 16
   Multiplying and Dividing Integers 17
   Order of Operations 18
   Ordering Integers and Numbers 19
   Integers and Absolute Value 20
   Factoring Numbers 21
   Greatest Common Factor 22
   Least Common Multiple 23
   Answers of Worksheets 24

Chapter 2 : Fractions and Decimals 27
   Simplifying Fractions 28
   Adding and Subtracting Fractions 29
   Multiplying and Dividing Fractions 30
   Adding and Subtracting Mixed Numbers 31
   Multiplying and Dividing Mixed Numbers 32
   Adding and Subtracting Decimals 33
   Multiplying and Dividing Decimals 34
   Comparing Decimals 35
   Rounding Decimals 36
   Answers of Worksheets 37

Chapter 3 : Proportions, Ratios, and Percent 40
   Simplifying Ratios 41
   Proportional Ratios 42
   Similarity and Ratios 43
   Ratio and Rates Word Problems 44

HiSET Subject Test – Mathematics

Percentage Calculations 45
Percent Problems 46
Discount, Tax and Tip 47
Percent of Change 48
Simple Interest 49
Answers of Worksheets 50

Chapter 4 : Exponents and Radicals Expressions 53

Multiplication Property of Exponents 54
Zero and Negative Exponents 55
Division Property of Exponents 56
Powers of Products and Quotients 57
Negative Exponents and Negative Bases 58
Scientific Notation 59
Square Roots 60
Simplifying Radical Expressions 61
Answers of Worksheets 62

Chapter 5 : Algebraic Expressions 65

Simplifying Variable Expressions 66
Simplifying Polynomial Expressions 67
Translate Phrases into an Algebraic Statement 68
The Distributive Property 69
Evaluating One Variable Expressions 70
Evaluating Two Variables Expressions 71
Combining like Terms 72
Answers of Worksheets 73

Chapter 6 : Equations and Inequalities 75

One–Step Equations 76
Multi–Step Equations 77
Graphing Single–Variable Inequalities 78
One–Step Inequalities 79
Multi-Step Inequalities 80
Systems of Equations 81
Systems of Equations Word Problems 82

HiSET Subject Test – Mathematics

    Answers of Worksheets 83

Chapter 7 : Linear Functions 87
    Finding Slope 88
    Graphing Lines Using Line Equation 89
    Writing Linear Equations 90
    Graphing Linear Inequalities 91
    Finding Midpoint 92
    Finding Distance of Two Points 93
    Answers of Worksheets 94

Chapter 8 : Polynomials 97
    Writing Polynomials in Standard Form 98
    Simplifying Polynomials 99
    Adding and Subtracting Polynomials 100
    Multiplying Monomials 101
    Multiplying and Dividing Monomials 102
    Multiplying a Polynomial and a Monomial 103
    Multiplying Binomials 104
    Factoring Trinomials 105
    Operations with Polynomials 106
    Answers of Worksheets 107

Chapter 9 : Functions Operations and Quadratic 111
    Evaluating Function 112
    Adding and Subtracting Functions 113
    Multiplying and Dividing Functions 114
    Composition of Functions 115
    Quadratic Equation 116
    Solving Quadratic Equations 117
    Quadratic Formula and the Discriminant 118
    Graphing Quadratic Functions 119
    Answers of Worksheets 120

Chapter 10 : Geometry and Solid Figures 123
    Angles 124
    Pythagorean Relationship 125

WWW.MathNotion.Com

HiSET Subject Test – Mathematics

Triangles...126

Polygons..127

Trapezoids...128

Circles..129

Cubes...130

Rectangular Prism...131

Cylinder...132

Pyramids and Cone..133

Answers of Worksheets...134

Chapter 11 : Statistics and Probability 137

Mean and Median..138

Mode and Range..139

Times Series..140

Stem–and–Leaf Plot..141

Pie Graph...142

Probability Problems...143

Answers of Worksheets...144

Chapter 12 : HiSET Test Review.. 147

Practice Test 1...151

Practice Test 2...163

Chapter 13 : Answers and Explanations.. 175

Answer Key...175

Score Your Test...177

Practice Test 1...179

Practice Test 2...189

HiSET Subject Test – Mathematics

Chapter 1:
Integers and Number Theory

Topics that you'll practice in this chapter:

- ✓ Rounding
- ✓ Whole Number Addition and Subtraction
- ✓ Whole Number Multiplication and Division
- ✓ Rounding and Estimates
- ✓ Adding and Subtracting Integers
- ✓ Multiplying and Dividing Integers
- ✓ Order of Operations
- ✓ Ordering Integers and Numbers
- ✓ Integers and Absolute Value
- ✓ Factoring Numbers
- ✓ Greatest Common Factor (GCF)
- ✓ Least Common Multiple (LCM)

"Wherever there is number, there is beauty." –Proclus

Rounding

✎ **Round each number to the nearest ten.**

1) 42 = ___   5) 19 = ___   9) 48 = ___

2) 88 = ___   6) 25 = ___   10) 81 = ___

3) 24 = ___   7) 93 = ___   11) 58 = ___

4) 57 = ___   8) 71 = ___   12) 87 = ___

✎ **Round each number to the nearest hundred.**

13) 198 = ___   17) 321 = ___   21) 580 = ___

14) 387 = ___   18) 433 = ___   22) 868 = ___

15) 816 = ___   19) 579 = ___   23) 480 = ___

16) 101 = ___   20) 825 = ___   24) 287 = ___

✎ **Round each number to the nearest thousand.**

25) 1,382 = ___   29) 9,099 = ___   33) 52,866 = ___

26) 3,420 = ___   30) 22,980 = ___   34) 85,190 = ___

27) 4,254 = ___   31) 45,188 = ___   35) 70,990 = ___

28) 6,861 = ___   32) 16,808 = ___   36) 26,869 = ___

HiSET Subject Test – Mathematics

Whole Number Addition and Subtraction

✎ Find the sum or subtract.

1) 1,240 + 658 = _____

2) 3,458 − 544 = _____

3) 2,259 − 752 = _____

4) 1,990 + 324 = _____

5) 3,088 + 229 = _____

6) 2,354 + 1,009 = _____

7) 2,855 + 4,455 = _____

8) 5,112 + 4,004 = _____

9) 4,822 − 2,007 = _____

10) 8,380 − 5,288 = _____

11) 3,227 + 4,150 = _____

12) 7,702 − 4,331 = _____

✎ Find the missing number.

13) 720 + ____ = 1,360

14) 2,115 − ____ = 1,103

15) ____ + 3,105 = 6,200

16) 5,250 − 3,280 = ____

17) 8,020 + ____ = 8,990

18) 7,302 − 4,700 = ____

WWW.MathNotion.Com

HiSET Subject Test – Mathematics

Whole Number Multiplication and Division

✍ **Calculate each product.**

1) 42 × 13

2) 70 × 15

3) 40 × 14

4) 22 × 20

5) 110 × 11

6) 150 × 16

✍ **Find the missing quotient.**

7) 564 ÷ 6 = _____

8) 270 ÷ 3 = _____

9) 640 ÷ 8 = _____

10) 450 ÷ 9 = _____

11) 112 ÷ 7 = _____

12) 1,260 ÷ 9 = _____

13) 3,000 ÷ 10 = _____

14) 2,400 ÷ 8 = _____

15) 3,200 ÷ 40 = _____

16) 6,300 ÷ 90 = _____

✍ **Calculate each problem.**

17) 400 ÷ 5 = N, N = __

18) 2,500 ÷ 10 = N, N = __

19) N ÷ 3 = 150, N = __

20) 42 × N = 252, N = __

21) 660 ÷ N = 330, N = __

22) N × 8 = 336, N = __

WWW.MathNotion.Com

HiSET Subject Test – Mathematics

Rounding and Estimates

✎ **Estimate the sum by rounding each number to the nearest ten.**

1) $13 + 22 = $ _____

2) $71 + 23 = $ _____

3) $61 + 58 = $ _____

4) $56 + 85 = $ _____

5) $368 + 249 = $ _____

6) $330 + 903 = $ _____

7) $471 + 293 = $ _____

8) $1,950 + 2,655 = $ _____

✎ **Estimate the product by rounding each number to the nearest ten.**

9) $32 \times 71 = $ _____

10) $12 \times 33 = $ _____

11) $31 \times 83 = $ _____

12) $19 \times 11 = $ _____

13) $42 \times 76 = $ _____

14) $63 \times 34 = $ _____

15) $19 \times 31 = $ _____

16) $59 \times 71 = $ _____

✎ **Estimate the sum or product by rounding each number to the nearest ten.**

17) $\begin{array}{r} 29 \\ \times\ 12 \\ \hline \end{array}$

18) $\begin{array}{r} 37 \\ \times\ 26 \\ \hline \end{array}$

19) $\begin{array}{r} 48 \\ +\ 82 \\ \hline \end{array}$

20) $\begin{array}{r} 65 \\ +44 \\ \hline \end{array}$

21) $\begin{array}{r} 37 \\ \times\ 14 \\ \hline \end{array}$

22) $\begin{array}{r} 71 \\ +\ 32 \\ \hline \end{array}$

WWW.MathNotion.Com

HiSET Subject Test – Mathematics

Adding and Subtracting Integers

✏ **Find each sum.**

1) $14 + (-6) =$

2) $(-13) + (-20) =$

3) $5 + (-28) =$

4) $50 + (-12) =$

5) $(-7) + (-15) + 3 =$

6) $30 + (-14) + 8 =$

7) $40 + (-10) + (-14) + 17 =$

8) $(-15) + (-20) + 13 + 35 =$

9) $40 + (-20) + (38 - 29) =$

10) $28 + (-12) + (30 - 12) =$

✏ **Find each difference.**

11) $(-18) - (-7) =$

12) $25 - (-14) =$

13) $(-20) - 36 =$

14) $34 - (-19) =$

15) $51 - (30 - 21) =$

16) $17 - (5) - (-24) =$

17) $(35 + 20) - (-46) =$

18) $48 - 16 - (-8) =$

19) $62 - (28 + 17) - (-15) =$

20) $58 - (-23) - (-31) =$

21) $19 - (-8) - (-13) =$

22) $(19 - 24) - (-14) =$

23) $27 - 33 - (-21) =$

24) $58 - (32 + 24) - (-9) =$

25) $36 - (-30) + (-17) =$

26) $27 - (-42) + (-31) =$

WWW.MathNotion.Com

HiSET Subject Test – Mathematics

Multiplying and Dividing Integers

✏️ **Find each product.**

1) $(-9) \times (-5) =$

2) $(-3) \times 9 =$

3) $8 \times (-12) =$

4) $(-7) \times (-20) =$

5) $(-3) \times (-5) \times 6 =$

6) $(14 - 3) \times (-8) =$

7) $12 \times (-9) \times (-3) =$

8) $(140 + 10) \times (-2) =$

9) $10 \times (-12 + 8) \times 3 =$

10) $(-8) \times (-5) \times (-10) =$

✏️ **Find each quotient.**

11) $42 \div (-7) =$

12) $(-48) \div (-6) =$

13) $(-40) \div (-8) =$

14) $54 \div (-2) =$

15) $152 \div 19 =$

16) $(-144) \div (-12) =$

17) $180 \div (-10) =$

18) $(-312) \div (-12) =$

19) $221 \div (-13) =$

20) $(-126) \div (6) =$

21) $(-161) \div (-7) =$

22) $-266 \div (-14) =$

23) $(-120) \div (-4) =$

24) $270 \div (-18) =$

25) $(-208) \div (-8) =$

26) $(135) \div (-15) =$

WWW.MathNotion.Com

HiSET Subject Test – Mathematics

Order of Operations

✎ **Evaluate each expression.**

1) $7 + (5 \times 4) =$

2) $14 - (3 \times 6) =$

3) $(19 \times 4) + 16 =$

4) $(16 - 7) - (8 \times 2) =$

5) $27 + (18 \div 3) =$

6) $(18 \times 8) \div 6 =$

7) $(32 \div 4) \times (-2) =$

8) $(9 \times 4) + (32 - 18) =$

9) $24 + (4 \times 3) + 7 =$

10) $(36 \times 3) \div (2 + 2) =$

11) $(-7) + (12 \times 3) + 11 =$

12) $(8 \times 5) - (24 \div 6) =$

13) $(7 \times 6 \div 3) - (12 + 9) =$

14) $(13 + 5 - 14) \times 3 - 2 =$

15) $(20 - 14 + 30) \times (64 \div 4) =$

16) $32 + (28 - (36 \div 9)) =$

17) $(7 + 6 - 4 - 7) + (15 \div 5) =$

18) $(85 - 20) + (20 - 18 + 7) =$

19) $(20 \times 2) + (14 \times 3) - 22 =$

20) $18 + 5 - (30 \times 3) + 20 =$

WWW.MathNotion.Com

HiSET Subject Test – Mathematics

Ordering Integers and Numbers

✎ **Order each set of integers from least to greatest.**

1) $8, -10, -5, -3, 4$ ___, ___, ___, ___, ___, ___

2) $-10, -18, 6, 14, 27$ ___, ___, ___, ___, ___, ___

3) $15, -8, -21, 21, -23$ ___, ___, ___, ___, ___, ___

4) $-14, -40, 23, -12, 47$ ___, ___, ___, ___, ___, ___

5) $59, -54, 32, -57, 36$ ___, ___, ___, ___, ___, ___

6) $68, 26, -19, 47, -34$ ___, ___, ___, ___, ___, ___

✎ **Order each set of integers from greatest to least.**

7) $18, 36, -16, -18, -10$ ___, ___, ___, ___, ___, ___

8) $27, 34, -12, -24, 94$ ___, ___, ___, ___, ___, ___

9) $50, -21, -13, 42, -2$ ___, ___, ___, ___, ___, ___

10) $37, 46, -20, -16, 86$ ___, ___, ___, ___, ___, ___

11) $-18, 88, -26, -59, 75$ ___, ___, ___, ___, ___, ___

12) $-65, -30, -25, 3, 14$ ___, ___, ___, ___, ___, ___

WWW.MathNotion.Com

HiSET Subject Test – Mathematics

Integers and Absolute Value

✎ **Write absolute value of each number.**

1) $|-2| =$

2) $|-27| =$

3) $|-20| =$

4) $|14| =$

5) $|6| =$

6) $|-55| =$

7) $|16| =$

8) $|2| =$

9) $|54| =$

10) $|-4| =$

11) $|-11|$

12) $|88| =$

13) $|0| =$

14) $|79| =$

15) $|-32| =$

16) $|-17| =$

17) $|42| =$

18) $|-46| =$

19) $|1| =$

20) $|-40| =$

✎ **Evaluate the value.**

21) $|-5| - \frac{|-21|}{7} =$

22) $14 - |3 - 15| - |-4| =$

23) $\frac{|-32|}{4} \times |-4| =$

24) $\frac{|7 \times (-3)|}{7} \times \frac{|-19|}{3} =$

25) $|4 \times (-5)| + \frac{|-40|}{5} =$

26) $\frac{|-45|}{9} \times \frac{|-24|}{12} =$

27) $|-12 + 8| \times \frac{|-7 \times 7|}{7} =$

28) $\frac{|-11 \times 2|}{4} \times |-16| =$

HiSET Subject Test – Mathematics

Factoring Numbers

✎ **List all positive factors of each number.**

1) 9

2) 16

3) 24

4) 30

5) 26

6) 46

7) 20

8) 68

9) 28

10) 98

11) 14

12) 54

13) 55

14) 18

15) 63

16) 34

17) 50

18) 62

19) 95

20) 64

21) 70

22) 45

23) 22

24) 65

HiSET Subject Test – Mathematics

Greatest Common Factor

✎ **Find the GCF for each number pair.**

1) 6, 2

2) 4, 5

3) 3, 12

4) 7, 3

5) 5, 10

6) 8, 48

7) 6, 18

8) 9, 15

9) 12, 18

10) 4, 36

11) 6, 10

12) 28, 52

13) 25, 10

14) 22, 24

15) 9, 54

16) 8, 54

17) 42, 14

18) 16, 40

19) 9, 2, 3

20) 5, 15, 10

21) 7, 9, 2

22) 16, 64

23) 30, 48

24) 36, 63

WWW.MathNotion.Com

HiSET Subject Test – Mathematics

Least Common Multiple

✎ Find the LCM for each number pair.

1) 6, 9

2) 15, 45

3) 16, 40

4) 12, 36

5) 18, 27

6) 14, 42

7) 6, 30

8) 8, 56

9) 7, 21

10) 8, 20

11) 15, 25

12) 7, 9

13) 4, 11

14) 8, 28

15) 28, 56

16) 40, 50

17) 12, 13

18) 22, 11

19) 36, 20

20) 15, 35

21) 18, 81

22) 30, 54

23) 18, 45

24) 75, 25

HiSET Subject Test – Mathematics

Answers of Worksheets

Rounding

1) 40
2) 90
3) 20
4) 60
5) 20
6) 30
7) 90
8) 70
9) 50
10) 80
11) 60
12) 90
13) 200
14) 400
15) 800
16) 100
17) 300
18) 400
19) 600
20) 800
21) 600
22) 900
23) 500
24) 300
25) 1,000
26) 3,000
27) 4,000
28) 7,000
29) 9,000
30) 23,000
31) 45,000
32) 17,000
33) 53,000
34) 85,000
35) 71,000
36) 27,000

Whole Number Addition and Subtraction

1) 1,898
2) 2,914
3) 1,507
4) 2,314
5) 3,317
6) 3,363
7) 7,310
8) 9,116
9) 2,815
10) 3,092
11) 7,377
12) 3,371
13) 640
14) 1,012
15) 3,095
16) 1,970
17) 970
18) 2,602

Whole Number Multiplication and Division

1) 546
2) 1,050
3) 560
4) 440
5) 1,210
6) 2,400
7) 94
8) 90
9) 80
10) 50
11) 16
12) 140
13) 300
14) 300
15) 80
16) 70
17) 80
18) 250
19) 450
20) 6
21) 2
22) 42

Rounding and Estimates

1) 30
2) 90
3) 120
4) 150
5) 620
6) 1,230
7) 760
8) 4,610
9) 2,100
10) 300
11) 2,400
12) 200
13) 3,200
14) 1,800
15) 600
16) 4,200
17) 300
18) 1,200
19) 130
20) 110
21) 400
22) 100

WWW.MathNotion.Com

HiSET Subject Test – Mathematics

Adding and Subtracting Integers

1) 8
2) −33
3) −23
4) 38
5) −19
6) 24
7) 33
8) 13
9) 29
10) 34
11) −11
12) 39
13) −56
14) 53
15) 42
16) 36
17) 101
18) 40
19) 32
20) 112
21) 40
22) 9
23) 15
24) 11
25) 49
26) 38

Multiplying and Dividing Integers

1) 45
2) −27
3) −96
4) 140
5) 90
6) −88
7) 324
8) −300
9) −120
10) −400
11) −6
12) 8
13) 5
14) −27
15) 8
16) 12
17) −18
18) 26
19) −17
20) −21
21) 23
22) 19
23) 30
24) −15
25) 26
26) −9

Order of Operations

1) 27
2) −4
3) 92
4) −7
5) 33
6) 24
7) −16
8) 50
9) 43
10) 27
11) 40
12) 36
13) −7
14) 10
15) 576
16) 56
17) 5
18) 74
19) 60
20) −47

Ordering Integers and Numbers

1) −10, −5, −3, 4, 8
2) −18, −10, 6, 14, 27
3) −23, −21, −8, 15, 21
4) −40, −14, −12, 23, 47
5) −57, −54, 32, 36, 59
6) −34, −19, 26, 47, 68
7) 36, 18, −10, −16, −18
8) 94, 34, 27, −12, −24
9) 50, 42, −2, −13, −21
10) 86, 46, 37, −16, −20
11) 88, 75, −18, −26, −59
12) 14, 3, −25, −30, −65

Integers and Absolute Value

1) 2
2) 27
3) 20
4) 14

WWW.MathNotion.Com

HiSET Subject Test – Mathematics

5) 6	11) 11	17) 42	23) 32
6) 55	12) 88	18) 46	24) 19
7) 16	13) 0	19) 1	25) 28
8) 2	14) 79	20) 40	26) 10
9) 54	15) 32	21) 2	27) 28
10) 4	16) 17	22) −2	28) 88

Factoring Numbers

1) 1, 3, 9	10) 1, 2, 7, 14, 49, 98	19) 1, 5, 19, 95
2) 1, 2, 4, 8, 16	11) 1, 2, 7, 14	20) 1, 2, 4, 8, 16, 32, 64
3) 1, 2, 3, 4, 6, 8, 12, 24	12) 1, 2, 3, 6, 9, 18, 27, 54	21) 1, 2, 5, 7, 10, 14, 35, 70
4) 1, 2, 3, 5, 6, 10, 15, 30	13) 1, 5, 11, 55	22) 1, 3, 5, 9, 15, 45
5) 1, 2, 13, 26	14) 1, 2, 3, 6, 9, 18	23) 1, 2, 11, 22
6) 1, 2, 23, 46	15) 1, 3, 7, 9, 21, 63	24) 1, 5, 13, 65
7) 1, 2, 4, 5, 10, 20	16) 1, 2, 17, 34	
8) 1, 2, 4, 17, 34, 68	17) 1, 2, 5, 10, 25, 50	
9) 1, 2, 4, 7, 14, 28	18) 1, 2, 31, 62	

Greatest Common Factor

1) 2	7) 6	13) 5	19) 1
2) 1	8) 3	14) 2	20) 5
3) 3	9) 6	15) 9	21) 1
4) 1	10) 4	16) 2	22) 16
5) 5	11) 2	17) 14	23) 6
6) 8	12) 4	18) 8	24) 9

Least Common Multiple

1) 18	7) 30	13) 44	19) 180
2) 45	8) 56	14) 56	20) 105
3) 80	9) 21	15) 56	21) 162
4) 36	10) 40	16) 200	22) 270
5) 54	11) 75	17) 156	23) 90
6) 42	12) 63	18) 22	24) 75

HiSET Subject Test – Mathematics

Chapter 2 :
Fractions and Decimals

Topics that you'll practice in this chapter:

- ✓ Simplifying Fractions
- ✓ Adding and Subtracting Fractions
- ✓ Multiplying and Dividing Fractions
- ✓ Adding and Subtract Mixed Numbers
- ✓ Multiplying and Dividing Mixed Numbers
- ✓ Adding and Subtracting Decimals
- ✓ Multiplying and Dividing Decimals
- ✓ Comparing Decimals
- ✓ Rounding Decimals

"A Man is like a fraction whose numerator is what he is and whose denominator is what he thinks of himself. The larger the denominator, the smaller the fraction." –Tolstoy

HiSET Subject Test – Mathematics

Simplifying Fractions

✎ Simplify each fraction to its lowest terms.

1) $\dfrac{5}{10} =$

2) $\dfrac{28}{35} =$

3) $\dfrac{27}{36} =$

4) $\dfrac{40}{80} =$

5) $\dfrac{14}{56} =$

6) $\dfrac{32}{48} =$

7) $\dfrac{52}{65} =$

8) $\dfrac{15}{60} =$

9) $\dfrac{80}{160} =$

10) $\dfrac{55}{77} =$

11) $\dfrac{28}{112} =$

12) $\dfrac{32}{64} =$

13) $\dfrac{63}{72} =$

14) $\dfrac{81}{90} =$

15) $\dfrac{35}{105} =$

16) $\dfrac{25}{70} =$

17) $\dfrac{80}{280} =$

18) $\dfrac{12}{81} =$

19) $\dfrac{36}{186} =$

20) $\dfrac{240}{540} =$

21) $\dfrac{70}{560} =$

✎ Find the answer for each problem.

22) Which of the following fractions equal to $\dfrac{3}{4}$? ____

    A. $\dfrac{60}{90}$ B. $\dfrac{43}{104}$ C. $\dfrac{48}{64}$ D. $\dfrac{150}{300}$

23) Which of the following fractions equal to $\dfrac{5}{8}$? ____

    A. $\dfrac{125}{200}$ B. $\dfrac{115}{200}$ C. $\dfrac{50}{100}$ D. $\dfrac{30}{90}$

24) Which of the following fractions equal to $\dfrac{3}{7}$? ____

    A. $\dfrac{58}{116}$ B. $\dfrac{54}{126}$ C. $\dfrac{270}{167}$ D. $\dfrac{42}{63}$

HiSET Subject Test – Mathematics

Adding and Subtracting Fractions

✏ **Find the sum.**

1) $\frac{5}{9} + \frac{4}{9} =$

2) $\frac{1}{2} + \frac{1}{7} =$

3) $\frac{3}{8} + \frac{1}{4} =$

4) $\frac{3}{5} + \frac{1}{2} =$

5) $\frac{1}{4} + \frac{3}{5} =$

6) $\frac{7}{8} + \frac{3}{8} =$

7) $\frac{1}{2} + \frac{7}{10} =$

8) $\frac{2}{5} + \frac{2}{3} =$

9) $\frac{5}{7} + \frac{2}{3} =$

10) $\frac{7}{12} + \frac{3}{4} =$

11) $\frac{5}{6} + \frac{2}{5} =$

12) $\frac{1}{12} + \frac{2}{3} =$

✏ **Find the difference.**

13) $\frac{1}{3} - \frac{1}{6} =$

14) $\frac{3}{4} - \frac{1}{8} =$

15) $\frac{1}{2} - \frac{1}{3} =$

16) $\frac{1}{4} - \frac{1}{5} =$

17) $\frac{5}{8} - \frac{2}{3} =$

18) $\frac{1}{4} - \frac{1}{7} =$

19) $\frac{5}{6} - \frac{1}{9} =$

20) $\frac{3}{4} - \frac{1}{6} =$

21) $\frac{7}{8} - \frac{1}{12} =$

22) $\frac{8}{15} - \frac{3}{5} =$

23) $\frac{3}{12} - \frac{1}{14} =$

24) $\frac{10}{13} - \frac{7}{26} =$

25) $\frac{6}{7} - \frac{3}{4} =$

26) $\frac{4}{5} - \frac{1}{8} =$

27) $\frac{4}{7} - \frac{2}{35} =$

28) $\frac{9}{16} - \frac{2}{8} =$

29) $\frac{8}{9} - \frac{7}{18} =$

30) $\frac{1}{2} - \frac{4}{9} =$

WWW.MathNotion.Com

HiSET Subject Test – Mathematics

Multiplying and Dividing Fractions

✎ Find the value of each expression in lowest terms.

1) $\dfrac{1}{5} \times \dfrac{15}{5} =$

2) $\dfrac{9}{12} \times \dfrac{4}{9} =$

3) $\dfrac{1}{16} \times \dfrac{8}{10} =$

4) $\dfrac{1}{24} \times \dfrac{8}{10} =$

5) $\dfrac{1}{5} \times \dfrac{1}{4} =$

6) $\dfrac{7}{9} \times \dfrac{1}{7} =$

7) $\dfrac{6}{7} \times \dfrac{1}{3} =$

8) $\dfrac{2}{8} \times \dfrac{2}{8} =$

9) $\dfrac{5}{8} \times \dfrac{3}{5} =$

10) $\dfrac{4}{7} \times \dfrac{1}{8} =$

11) $\dfrac{7}{15} \times \dfrac{5}{7} =$

12) $\dfrac{3}{10} \times \dfrac{5}{9} =$

✎ Find the value of each expression in lowest terms.

13) $\dfrac{1}{4} \div \dfrac{1}{8} =$

14) $\dfrac{1}{10} \div \dfrac{1}{5} =$

15) $\dfrac{3}{4} \div \dfrac{1}{5} =$

16) $\dfrac{1}{3} \div \dfrac{5}{6} =$

17) $\dfrac{1}{7} \div \dfrac{8}{42} =$

18) $\dfrac{3}{4} \div \dfrac{1}{6} =$

19) $\dfrac{2}{7} \div \dfrac{7}{13} =$

20) $\dfrac{1}{24} \div \dfrac{3}{16} =$

21) $\dfrac{7}{12} \div \dfrac{5}{6} =$

22) $\dfrac{22}{18} \div \dfrac{11}{9} =$

23) $\dfrac{9}{35} \div \dfrac{3}{7} =$

24) $\dfrac{2}{7} \div \dfrac{8}{21} =$

25) $\dfrac{1}{9} \div \dfrac{2}{5} =$

26) $\dfrac{5}{12} \div \dfrac{3}{5} =$

27) $\dfrac{3}{20} \div \dfrac{1}{6} =$

28) $\dfrac{8}{20} \div \dfrac{3}{4} =$

29) $\dfrac{5}{6} \div \dfrac{2}{9} =$

30) $\dfrac{5}{11} \div \dfrac{3}{4} =$

WWW.MathNotion.Com

HiSET Subject Test – Mathematics

Adding and Subtracting Mixed Numbers

✏️ Find the sum.

1) $3\frac{1}{3} + 2\frac{1}{6} =$

2) $4\frac{1}{2} + 3\frac{1}{2} =$

3) $3\frac{3}{8} + 1\frac{1}{8} =$

4) $2\frac{1}{4} + 2\frac{1}{3} =$

5) $3\frac{5}{6} + 2\frac{7}{12} =$

6) $5\frac{4}{15} + 3\frac{3}{5} =$

7) $2\frac{1}{3} + 4\frac{3}{7} =$

8) $3\frac{1}{2} + 4\frac{2}{5} =$

9) $5\frac{2}{5} + 6\frac{3}{7} =$

10) $8\frac{5}{16} + 6\frac{1}{12} =$

✏️ Find the difference.

11) $3\frac{1}{4} - 1\frac{3}{4} =$

12) $6\frac{3}{5} - 4\frac{2}{5} =$

13) $4\frac{1}{3} - 3\frac{1}{9} =$

14) $7\frac{1}{7} - 5\frac{1}{2} =$

15) $5\frac{1}{3} - 2\frac{1}{12} =$

16) $8\frac{1}{5} - 4\frac{1}{3} =$

17) $9\frac{1}{4} - 6\frac{1}{8} =$

18) $11\frac{7}{15} - 8\frac{3}{5} =$

19) $14\frac{5}{6} - 11\frac{3}{5} =$

20) $18\frac{2}{7} - 14\frac{1}{5} =$

21) $9\frac{1}{3} - 4\frac{1}{4} =$

22) $6\frac{1}{8} - 4\frac{1}{16} =$

23) $19\frac{3}{8} - 15\frac{1}{3} =$

24) $11\frac{1}{9} - 8\frac{1}{8} =$

25) $17\frac{1}{7} - 11\frac{1}{5} =$

26) $16\frac{2}{9} - 9\frac{5}{7} =$

WWW.MathNotion.Com

HiSET Subject Test – Mathematics

Multiplying and Dividing Mixed Numbers

✎ **Find the product.**

1) $5\frac{1}{2} \times 2\frac{1}{4} =$

2) $5\frac{1}{3} \times 4\frac{1}{3} =$

3) $5\frac{3}{4} \times 6\frac{1}{4} =$

4) $3\frac{1}{3} \times 2\frac{3}{5} =$

5) $4\frac{8}{10} \times 1\frac{1}{24} =$

6) $6\frac{2}{7} \times 1\frac{1}{11} =$

7) $8\frac{2}{3} \times 3\frac{1}{2} =$

8) $3\frac{4}{7} \times 2\frac{1}{5} =$

9) $5\frac{2}{8} \times 4\frac{1}{6} =$

10) $7\frac{3}{3} \times 1\frac{3}{8} =$

✎ **Find the quotient.**

11) $2\frac{2}{5} \div 4\frac{1}{5} =$

12) $4\frac{1}{6} \div 3\frac{1}{3} =$

13) $6\frac{1}{3} \div 1\frac{1}{2} =$

14) $7\frac{1}{10} \div 2\frac{2}{5} =$

15) $3\frac{1}{3} \div 1\frac{1}{9} =$

16) $1\frac{1}{10} \div 4\frac{1}{2} =$

17) $1\frac{3}{16} \div 5\frac{1}{4} =$

18) $4\frac{1}{3} \div 4\frac{3}{4} =$

19) $9\frac{1}{3} \div 2\frac{1}{4} =$

20) $15\frac{1}{3} \div 5\frac{1}{2} =$

21) $4\frac{1}{6} \div 1\frac{1}{5} =$

22) $1\frac{1}{18} \div 1\frac{2}{9} =$

23) $4\frac{2}{7} \div 1\frac{3}{10} =$

24) $7\frac{1}{3} \div 2\frac{2}{11} =$

25) $8\frac{2}{5} \div 1\frac{1}{6} =$

26) $9\frac{1}{3} \div 2\frac{1}{7} =$

WWW.MathNotion.Com

HiSET Subject Test – Mathematics

Adding and Subtracting Decimals

✏ **Add and subtract decimals.**

1) 35.19 − 24.28 = _____

4) 38.72 − 21.68 = _____

7) 86.09 − 35.14 = _____

2) 34.29 + 42.58 = _____

5) 57.39 + 26.54 = _____

8) 54.51 + 32.66 = _____

3) 61.20 + 33.75 = _____

6) 70.24 − 42.35 = _____

9) 114.21 − 88.69 = _____

✏ **Find the missing number.**

10) ___ + 2.8 = 5.4

11) 4.1 + ___ = 5.88

12) 6.45 + ___ = 8

13) 7.25 − ___ = 3.40

14) ___ − 2.35 = 4.25

15) ___ − 19.85 = 6.54

16) 22.15 + ___ = 28.95

17) ___ − 37.16 = 9.42

18) ___ + 24.50 = 34.19

19) 72.40 + ___ = 125.20

WWW.MathNotion.Com

HiSET Subject Test – Mathematics

Multiplying and Dividing Decimals

✏ **Find the product.**

1) $0.5 \times 0.6 =$

2) $3.3 \times 0.4 =$

3) $1.28 \times 0.5 =$

4) $0.35 \times 0.6 =$

5) $1.85 \times 0.6 =$

6) $0.24 \times 0.5 =$

7) $5.25 \times 1.4 =$

8) $18.5 \times 4.6 =$

9) $15.4 \times 6.8 =$

10) $19.5 \times 2.6 =$

11) $32.2 \times 1.5 =$

12) $78.4 \times 4.5 =$

✏ **Find the quotient.**

13) $1.85 \div 10 =$

14) $74.6 \div 100 =$

15) $3.6 \div 3 =$

16) $9.6 \div 0.4 =$

17) $15.5 \div 0.5 =$

18) $32.8 \div 0.2 =$

19) $22.15 \div 1{,}000 =$

20) $53.55 \div 0.7 =$

21) $322.2 \div 0.2 =$

22) $50.67 \div 0.18 =$

23) $77.4 \div 0.8 =$

24) $27.93 \div 0.03 =$

HiSET Subject Test – Mathematics

Comparing Decimals

✎ **Write the correct comparison symbol (>, < or =).**

1) 0.70 ☐ 0.070

2) 0.049 ☐ 0.49

3) 5.090 ☐ 5.09

4) 2.57 ☐ 2.05

5) 9.03 ☐ 0.930

6) 6.06 ☐ 6.6

7) 7.02 ☐ 7.020

8) 3.04 ☐ 3.2

9) 3.61 ☐ 3.245

10) 0.986 ☐ 0.0986

11) 17.24 ☐ 17.240

12) 0.759 ☐ 0.81

13) 9.040 ☐ 9.40

14) 5.73 ☐ 5.213

15) 9.44 ☐ 9.404

16) 7.17 ☐ 7.170

17) 4.85 ☐ 4.085

18) 9.041 ☐ 9.40

19) 3.033 ☐ 3.030

20) 4.97 ☐ 4.970

WWW.MathNotion.Com

HiSET Subject Test – Mathematics

Rounding Decimals

✎ **Round each decimal to the nearest whole number.**

1) 28.12   3) 16.22   5) 7.95

2) 6.9   4) 8.5   6) 52.7

✎ **Round each decimal to the nearest tenth.**

7) 31.761   9) 94.729   11) 13.219

8) 14.421   10) 77.89   12) 59.89

✎ **Round each decimal to the nearest hundredth.**

13) 8.428   15) 55.3786   17) 62.241

14) 23.812   16) 231.912   18) 19.447

✎ **Round each decimal to the nearest thousandth.**

19) 15.54324   21) 243.8652   23) 67.1983

20) 34.62586   22) 80.4529   24) 72.36788

WWW.MathNotion.Com

HiSET Subject Test – Mathematics

Answers of Worksheets

Simplifying Fractions

1) $\frac{1}{2}$ 7) $\frac{4}{5}$ 13) $\frac{7}{8}$ 19) $\frac{6}{31}$

2) $\frac{4}{5}$ 8) $\frac{1}{4}$ 14) $\frac{9}{10}$ 20) $\frac{4}{9}$

3) $\frac{3}{4}$ 9) $\frac{1}{2}$ 15) $\frac{1}{3}$ 21) $\frac{1}{8}$

4) $\frac{1}{2}$ 10) $\frac{5}{7}$ 16) $\frac{5}{14}$ 22) C

5) $\frac{1}{4}$ 11) $\frac{1}{4}$ 17) $\frac{2}{7}$ 23) A

6) $\frac{2}{3}$ 12) $\frac{1}{2}$ 18) $\frac{4}{27}$ 24) B

Adding and Subtracting Fractions

1) $\frac{9}{9} = 1$ 9) $1\frac{8}{21}$ 17) $-\frac{1}{24}$ 25) $\frac{3}{28}$

2) $\frac{9}{14}$ 10) $1\frac{1}{3}$ 18) $\frac{3}{28}$ 26) $\frac{27}{40}$

3) $\frac{5}{8}$ 11) $1\frac{7}{30}$ 19) $\frac{13}{18}$ 27) $\frac{18}{35}$

4) $1\frac{1}{10}$ 12) $\frac{3}{4}$ 20) $\frac{7}{12}$ 28) $\frac{5}{16}$

5) $\frac{17}{20}$ 13) $\frac{1}{6}$ 21) $\frac{19}{24}$ 29) $\frac{1}{2}$

6) $1\frac{1}{4}$ 14) $\frac{5}{8}$ 22) $-\frac{1}{15}$ 30) $\frac{1}{18}$

7) $1\frac{1}{5}$ 15) $\frac{1}{6}$ 23) $\frac{5}{28}$

8) $1\frac{1}{15}$ 16) $\frac{1}{20}$ 24) $\frac{1}{2}$

Multiplying and Dividing Fractions

1) $\frac{3}{5}$ 4) $\frac{1}{30}$ 7) $\frac{2}{7}$ 11) $\frac{1}{3}$

2) $\frac{1}{3}$ 5) $\frac{1}{20}$ 8) $\frac{1}{16}$ 12) $\frac{1}{6}$

3) $\frac{1}{20}$ 6) $\frac{1}{9}$ 9) $\frac{3}{8}$ 13) 2

    10) $\frac{1}{14}$ 14) $\frac{1}{2}$

WWW.MathNotion.Com

HiSET Subject Test – Mathematics

15) $3\frac{3}{4}$
16) $\frac{2}{5}$
17) $\frac{3}{4}$
18) $4\frac{1}{2}$

19) $\frac{26}{49}$
20) $\frac{2}{9}$
21) $\frac{7}{10}$
22) 1

23) $\frac{3}{5}$
24) $\frac{3}{4}$
25) $\frac{5}{18}$
26) $\frac{25}{36}$

27) $\frac{9}{10}$
28) $\frac{8}{15}$
29) $3\frac{3}{4}$
30) $\frac{20}{33}$

Adding and Subtracting Mixed Numbers

1) $5\frac{1}{2}$
2) 8
3) $4\frac{1}{2}$
4) $4\frac{7}{12}$
5) $6\frac{5}{12}$
6) $8\frac{13}{15}$
7) $6\frac{16}{21}$

8) $7\frac{9}{10}$
9) $11\frac{29}{35}$
10) $14\frac{19}{48}$
11) $1\frac{1}{2}$
12) $2\frac{1}{5}$
13) $1\frac{2}{9}$
14) $1\frac{9}{14}$

15) $3\frac{1}{4}$
16) $3\frac{13}{15}$
17) $3\frac{1}{8}$
18) $2\frac{13}{15}$
19) $3\frac{7}{30}$
20) $4\frac{3}{35}$
21) $5\frac{1}{12}$

22) $2\frac{1}{16}$
23) $4\frac{1}{24}$
24) $2\frac{71}{72}$
25) $5\frac{33}{35}$
26) $6\frac{32}{63}$

Multiplying and Dividing Mixed Numbers

1) $12\frac{3}{8}$
2) $23\frac{1}{9}$
3) $35\frac{15}{16}$
4) $8\frac{2}{3}$
5) 5
6) $6\frac{6}{7}$
7) $30\frac{1}{3}$
8) $7\frac{6}{7}$
9) $21\frac{7}{8}$

10) 11
11) $\frac{4}{7}$
12) $1\frac{1}{4}$
13) $4\frac{2}{9}$
14) $2\frac{23}{24}$
15) 3
16) $\frac{11}{45}$
17) $\frac{19}{84}$
18) $\frac{52}{57}$

19) $4\frac{4}{27}$
20) $2\frac{26}{33}$
21) $3\frac{17}{36}$
22) $\frac{19}{22}$
23) $3\frac{27}{91}$
24) $3\frac{13}{36}$
25) $7\frac{1}{5}$
26) $4\frac{16}{45}$

Adding and Subtracting Decimals

1) 10.91
2) 76.87
3) 94.95
4) 17.04

WWW.MathNotion.Com

HiSET Subject Test – Mathematics

5) 83.93
6) 27.89
7) 50.95
8) 87.17

9) 25.52
10) 2.6
11) 1.78
12) 1.55

13) 3.85
14) 6.6
15) 26.39
16) 6.8

17) 46.58
18) 9.69
19) 52.8

Multiplying and Dividing Decimals

1) 0.3
2) 1.32
3) 0.64
4) 0.21
5) 1.11
6) 0.12

7) 7.35
8) 85.1
9) 104.72
10) 50.7
11) 48.3
12) 352.8

13) 0.185
14) 0.746
15) 1.2
16) 24
17) 31
18) 164

19) 0.02215
20) 76.5
21) 1,611
22) 281.5
23) 96.75
24) 931

Comparing Decimals

1) >
2) <
3) =
4) >
5) >

6) <
7) =
8) <
9) >
10) >

11) =
12) <
13) <
14) >
15) >

16) =
17) >
18) <
19) >
20) =

Rounding Decimals

1) 28
2) 7
3) 16
4) 9
5) 8
6) 53
7) 31.8
8) 14.4

9) 94.7
10) 77.9
11) 13.2
12) 59.9
13) 8.43
14) 23.81
15) 55.38
16) 231.91

17) 62.24
18) 19.45
19) 15.543
20) 34.626
21) 243.865
22) 80.453
23) 67.198
24) 72.368

WWW.MathNotion.Com

HiSET Subject Test – Mathematics

Chapter 3 : Proportions, Ratios, and Percent

Topics that you'll practice in this chapter:

- ✓ Simplifying Ratios
- ✓ Proportional Ratios
- ✓ Similarity and Ratios
- ✓ Ratio and Rates Word Problems
- ✓ Percentage Calculations
- ✓ Percent Problems
- ✓ Discount, Tax and Tip
- ✓ Percent of Change
- ✓ Simple Interest

Without mathematics, there's nothing you can do. Everything around you is mathematics. Everything around you is numbers." – Shakuntala Devi

HiSET Subject Test – Mathematics

Simplifying Ratios

✎ **Reduce each ratio.**

1) $15:20 = $ ___ : ___

2) $7:70 = $ ___ : ___

3) $16:28 = $ ___ : ___

4) $7:21 = $ ___ : ___

5) $4:40 = $ ___ : ___

6) $6:48 = $ ___ : ___

7) $16:64 = $ ___ : ___

8) $10:25 = $ ___ : ___

9) $8:48 = $ ___ : ___

10) $49:63 = $ ___ : ___

11) $18:27 = $ ___ : ___

12) $35:10 = $ ___ : ___

13) $90:9 = $ ___ : ___

14) $24:32 = $ ___ : ___

15) $7:56 = $ ___ : ___

16) $45:63 = $ ___ : ___

17) $56:72 = $ ___ : ___

18) $26:13 = $ ___ : ___

19) $15:45 = $ ___ : ___

20) $28:4 = $ ___ : ___

21) $24:48 = $ ___ : ___

22) $30:24 = $ ___ : ___

23) $70:140 = $ ___ : ___

24) $6:180 = $ ___ : ___

✎ **Write each ratio as a fraction in simplest form.**

25) $6:12 =$

26) $30:50 =$

27) $15:35 =$

28) $9:27 =$

29) $8:24 =$

30) $18:84 =$

31) $7:14 =$

32) $7:35 =$

33) $40:96 =$

34) $12:54 =$

35) $44:52 =$

36) $12:27 =$

37) $15:180 =$

38) $39:143 =$

39) $20:300 =$

40) $30:120 =$

41) $56:42 =$

42) $26:130 =$

43) $66:123 =$

44) $70:630 =$

45) $75:125 =$

HiSET Subject Test – Mathematics

Proportional Ratios

✎ **Fill in the blanks; Calculate each proportion.**

1) $3:8 = \_\_:48$ 7) $20:3 = \_\_:15$

2) $2:5 = 20:\_\_$ 8) $1:3 = \_\_:75$

3) $1:9 = \_\_:81$ 9) $7:6 = \_\_:60$

4) $6:7 = 12:\_\_$ 10) $8:5 = \_\_:45$

5) $9:2 = 63:\_\_$ 11) $3:10 = 60:\_\_$

6) $8:7 = \_\_:49$ 12) $6:11 = 42:\_\_$

✎ **State if each pair of ratios form a proportion.**

13) $\frac{3}{20}$ and $\frac{9}{60}$ 17) $\frac{1}{9}$ and $\frac{12}{81}$ 21) $\frac{6}{19}$ and $\frac{30}{85}$

14) $\frac{1}{7}$ and $\frac{6}{42}$ 18) $\frac{7}{8}$ and $\frac{21}{28}$ 22) $\frac{5}{9}$ and $\frac{40}{81}$

15) $\frac{3}{7}$ and $\frac{24}{56}$ 19) $\frac{9}{13}$ and $\frac{27}{39}$ 23) $\frac{9}{14}$ and $\frac{108}{168}$

16) $\frac{4}{9}$ and $\frac{12}{18}$ 20) $\frac{1}{8}$ and $\frac{8}{64}$ 24) $\frac{15}{23}$ and $\frac{360}{552}$

✎ **Calculate each proportion.**

25) $\frac{20}{25} = \frac{32}{x}$, $x = \_\_\_$ 29) $\frac{7}{9} = \frac{x}{81}$, $x = \_\_\_$ 33) $\frac{5}{8} = \frac{x}{88}$, $x = \_\_\_$

26) $\frac{1}{8} = \frac{32}{x}$, $x = \_\_\_$ 30) $\frac{1}{5} = \frac{13}{x}$, $x = \_\_\_$ 34) $\frac{4}{15} = \frac{x}{240}$, $x = \_\_\_$

27) $\frac{15}{5} = \frac{21}{x}$, $x = \_\_\_$ 31) $\frac{9}{5} = \frac{36}{x}$, $x = \_\_\_$ 35) $\frac{9}{19} = \frac{x}{266}$, $x = \_\_\_$

28) $\frac{1}{7} = \frac{x}{294}$, $x = \_\_\_$ 32) $\frac{6}{13} = \frac{48}{x}$, $x = \_\_\_$ 36) $\frac{7}{15} = \frac{x}{270}$, $x = \_\_\_$

WWW.MathNotion.Com

HiSET Subject Test – Mathematics

Similarity and Ratios

✎ **Each pair of figures is similar. Find the missing side.**

1)

2)

3)

4)

✎ **Calculate.**

5) Two rectangles are similar. The first is 24 feet wide and 120 feet long. The second is 30 feet wide. What is the length of the second rectangle? _____

6) Two rectangles are similar. One is 5 meters by 36 meters. The longer side of the second rectangle is 90 meters. What is the other side of the second rectangle? _____

7) A building casts a shadow 25 ft long. At the same time a girl 10 ft tall casts a shadow 5 ft long. How tall is the building? _____

8) The scale of a map of Texas is 4 inches: 32 miles. If you measure the distance from Dallas to Martin County as 38.4 inches, approximately how far is Martin County from Dallas? _____

WWW.MathNotion.Com

HiSET Subject Test – Mathematics

Ratio and Rates Word Problems

✎ **Find the answer for each word problem.**

1) Mason has 24 red cards and 36 green cards. What is the ratio of Mason 's red cards to his green cards? _____

2) In a party, 45 soft drinks are required for every 54 guests. If there are 378 guests, how many soft drinks is required? _____

3) In Mason's class, 42 of the students are tall and 24 are short. In Michael's class 84 students are tall and 48 students are short. Which class has a higher ratio of tall to short students? _____

4) The price of 5 apples at the Quick Market is $4.6. The price of 7 of the same apples at Walmart is $5.95. Which place is the better buy? _____

5) The bakers at a Bakery can make 90 bagels in 3 hours. How many bagels can they bake in 24 hours? What is that rate per hour? _____

6) You can buy 5 cans of green beans at a supermarket for $5.75. How much does it cost to buy 45 cans of green beans? _____

7) The ratio of boys to girls in a class is 4: 7. If there are 32 boys in the class, how many girls are in that class? _____

8) The ratio of red marbles to blue marbles in a bag is 3: 7. If there are 50 marbles in the bag, how many of the marbles are red? _____

WWW.MathNotion.Com

HiSET Subject Test – Mathematics

Percentage Calculations

✍ **Calculate the given percent of each value.**

1) 3% of 60 = ___

2) 20% of 32 = ___

3) 4% of 72 = ___

4) 16% of 32 = ___

5) 25% of 124 = ___

6) 35% of 56 = ___

7) 15% of 20 = ___

8) 14% of 150 = ___

9) 80% of 50 = ___

10) 12% of 115 = ___

11) 72% of 250 = ___

12) 52% of 500 = ___

13) 70% of 400 = ___

14) 27% of 145 = ___

15) 90% of 64 = ___

16) 60% of 55 = ___

17) 22% of 210 = ___

18) 8% of 235 = ___

✍ **Calculate the percent of each given value.**

19) ___% of 25 = 5

20) ___% of 40 = 20

21) ___% of 25 = 2

22) ___% of 50 = 16

23) ___% of 250 = 5

24) ___% of 40 = 32

25) ___% of 125 = 20

26) ___% of 700 = 49

27) ___% of 350 = 49

28) ___% of 500 = 210

✍ **Calculate each percent problem.**

29) A Cinema has 250 seats. 60 seats were sold for the current movie. What percent of seats are empty? ___ %

30) There are 68 boys and 92 girls in a class. 75% of the students in the class take the bus to school. How many students do not take the bus to school? ___

HiSET Subject Test – Mathematics

Percent Problems

✎ **Calculate each problem.**

1) 9 is what percent of 45? ___%

2) 60 is what percent of 120? ___%

3) 10 is what percent of 200? ___%

4) 15 is what percent of 125? ___%

5) 10 is what percent of 400? ___%

6) 66 is what percent of 55? ___%

7) 40 is what percent of 160? ___%

8) 40 is what percent of 50? ___%

9) 120 is what percent of 800? ___%

10) 78 is what percent of 120? ___%

11) 36 is what percent of 144? ___%

12) 17 is what percent of 85? ___%

13) 90 is what percent of 900? ___%

14) 36 is what percent of 16? ___%

15) 63 is what percent of 14? ___%

16) 18 is what percent of 60? ___%

17) 126 is what percent of 200? ___%

18) 232 is what percent of 40? ___%

✎ **Calculate each percent word problem.**

19) There are 40 employees in a company. On a certain day, 25 were present. What percent showed up for work? ____%

20) A metal bar weighs 60 ounces. 25% of the bar is gold. How many ounces of gold are in the bar? _____

21) A crew is made up of 12 women; the rest are men. If 15% of the crew are women, how many people are in the crew? _____

22) There are 40 students in a class and 8 of them are girls. What percent are boys? ____%

23) The Royals softball team played 400 games and won 280 of them. What percent of the games did they lose? ____%

HiSET Subject Test – Mathematics

Discount, Tax and Tip

✏ **Find the selling price of each item.**

1) Original price of a computer: $420
   Tax: 8% Selling price: $_____

2) Original price of a laptop: $280
   Tax: 4% Selling price: $_____

3) Original price of a sofa: $820
   Tax: 5% Selling price: $_____

4) Original price of a car: $15,800
   Tax: 3.6%   Selling price: $_____

5) Original price of a Table: $250
   Tax: 9% Selling price: $_____

6) Original price of a house: $630,000
   Tax: 1.8%   Selling price: $_____

7) Original price of a tablet: $450
   Discount: 30%   Selling price: $____

8) Original price of a chair: $390
   Discount: 8%   Selling price: $____

9) Original price of a book: $75
   Discount: 42%   Selling price: $____

10) Original price of a cellphone: $820
    Discount: 23%   Selling price: $____

11) Food bill: $45
    Tip: 15%   Price: $_____

12) Food bill: $32
    Tipp: 20%   Price: $_____

13) Food bill: $90
    Tip: 35%   Price: $_____

14) Food bill: $42
    Tipp: 12%   Price: $_____

✏ **Find the answer for each word problem.**

15) Nicolas hired a moving company. The company charged $500 for its services, and Nicolas gives the movers a 40% tip. How much does Nicolas tip the movers? $_____

16) Mason has lunch at a restaurant and the cost of his meal is $90. Mason wants to leave a 25% tip. What is Mason's total bill including tip? $_____

17) The sales tax in Texas is 19.80% and an item costs $350. How much is the tax? $_____

18) The price of a table at Best Buy is $680. If the sales tax is 5%, what is the final price of the table including tax? $_____

HiSET Subject Test – Mathematics

Percent of Change

✒ **Find each percent of change.**

1) From 150 to 450. ___ %

2) From 50 ft to 250 ft. ___ %

3) From $60 to $360. ___ %

4) From 60 cm to 180 cm. ___ %

5) From 15 to 45. ___ %

6) From 80 to 16. ___ %

7) From 120 to 360. ___ %

8) From 900 to 450. ___ %

9) From 1,000 to 200. ___ %

10) From 144 to 36. ___ %

✒ **Calculate each percent of change word problem.**

11) Bob got a raise, and his hourly wage increased from $42 to $63. What is the percent increase? ___ %

12) The price of a pair of shoes increases from $50 to $61. What is the percent increase? ___ %

13) At a coffee shop, the price of a cup of coffee increased from $4.80 to $5.76. What is the percent increase in the cost of the coffee? ___ %

14) 51 cm are cut from 85 cm board. What is the percent decrease in length? ___ %

15) In a class, the number of students has been increased from 54 to 81. What is the percent increase? ___ %

16) The price of gasoline rises from $24.40 to $30.50 in one month. By what percent did the gas price rise? ___ %

17) A shirt was originally priced at $38. It went on sale for $24.70. What was the percent that the shirt was discounted? ___ %

WWW.MathNotion.Com

HiSET Subject Test – Mathematics

Simple Interest

✍ **Determine the simple interest for these loans.**

1) $480 at 11% for 3 years. $ _____

2) $4,200 at 7% for 4 years. $ _____

3) $2,500 at 20% for 3 years. $ _____

4) $6,800 at 3.9% for 4 months. $ _____

5) $800 at 6% for 7 months. $ _____

6) $36,000 at 4.2% for 6 years. $ _____

7) $6,500 at 7% for 4 years. $ _____

8) $850 at 9.5% for 2 years. $ _____

9) $1,200 at 5.8% for 9 months. $ _____

10) $3,000 at 4.5% for 7 years. $ _____

✍ **Calculate each simple interest word problem.**

11) A new car, valued at $22,000, depreciates at 8.5% per year. What is the value of the car one year after purchase? $_____

12) Sara puts $9,000 into an investment yielding 6% annual simple interest; she left the money in for three years. How much interest does Sara get at the end of those three years? $_____

13) A bank is offering 12% simple interest on a savings account. If you deposit $16,400, how much interest will you earn in two years? $_____

14) $720 interest is earned on a principal of $6,000 at a simple interest rate of 4% interest per year. For how many years was the principal invested? _____

15) In how many years will $2,200 yield an interest of $440 at 4% simple interest? _____

16) Jim invested $8,000 in a bond at a yearly rate of 4.5%. He earned $1,440 in interest. How long was the money invested? _____

WWW.MathNotion.Com

HiSET Subject Test – Mathematics

Answers of Worksheets

Simplifying Ratios

1) 3 : 4
2) 1 : 10
3) 4 : 7
4) 1 : 3
5) 1 : 10
6) 1 : 8
7) 2 : 8
8) 2 : 5
9) 1 : 6
10) 7 : 9
11) 2 : 3
12) 7 : 2
13) 10 : 1
14) 3 : 4
15) 1 : 8
16) 5 : 7
17) 7 : 9
18) 2 : 1
19) 1 : 3
20) 7 : 1
21) 1 : 2
22) 5 : 4
23) 1 : 2
24) 1 : 30
25) $\frac{1}{2}$
26) $\frac{3}{5}$
27) $\frac{3}{7}$
28) $\frac{1}{3}$
29) $\frac{1}{3}$
30) $\frac{3}{14}$
31) $\frac{1}{2}$
32) $\frac{1}{5}$
33) $\frac{5}{12}$
34) $\frac{2}{9}$
35) $\frac{11}{13}$
36) $\frac{4}{9}$
37) $\frac{1}{12}$
38) $\frac{3}{11}$
39) $\frac{1}{15}$
40) $\frac{1}{4}$
41) $\frac{4}{3}$
42) $\frac{1}{5}$
43) $\frac{22}{41}$
44) $\frac{1}{9}$
45) $\frac{3}{5}$

Proportional Ratios

1) 18
2) 50
3) 9
4) 14
5) 14
6) 56
7) 100
8) 25
9) 70
10) 72
11) 200
12) 77
13) Yes
14) Yes
15) Yes
16) No
17) No
18) No
19) Yes
20) Yes
21) No
22) No
23) Yes
24) Yes
25) 40
26) 256
27) 7
28) 42
29) 63
30) 65
31) 20
32) 104
33) 55
34) 64
35) 126
36) 126

Similarity and ratios

1) 15
2) 5
3) 15
4) 13
5) 150 feet
6) 12.5 meters
7) 50 feet
8) 307.2 miles

Ratio and Rates Word Problems

1) 2 : 3
2) 315

WWW.MathNotion.Com

HiSET Subject Test – Mathematics

3) The ratio for both classes is 7 to 4.
4) Walmart is a better buy.
5) 720, the rate is 30 per hour.

6) $51.75
7) 56
8) 15

Percentage Calculations

1) 1.8
2) 6.4
3) 2.88
4) 5.12
5) 31
6) 19.6
7) 3
8) 21
9) 40
10) 13.8

11) 180
12) 260
13) 280
14) 39.15
15) 57.6
16) 33
17) 46.2
18) 18.8
19) 20%
20) 50%

21) 8%
22) 32%
23) 2%
24) 80%
25) 16%
26) 7%
27) 14%
28) 42%
29) 76%
30) 40

Percent Problems

1) 20%
2) 50%
3) 5%
4) 12%
5) 2.5%
6) 120%
7) 25%
8) 80%

9) 15%
10) 65%
11) 25%
12) 20%
13) 10%
14) 225%
15) 450%
16) 30%

17) 63%
18) 580%
19) 62.5%
20) 15 ounces
21) 80
22) 80%
23) 30%

Discount, Tax and Tip

1) $453.60
2) $291.20
3) $861.00
4) $16,368.80
5) $272.50
6) $641,340

7) $315.00
8) $358.80
9) $43.50
10) $631.40
11) $51.75
12) $38.40

13) $121.50
14) $47.04
15) $200.00
16) $112.50
17) $69.30
18) $714.00

HiSET Subject Test – Mathematics

Percent of Change

1) 200%   7) 200%   13) 20%
2) 400%   8) 50%   14) 60%
3) 500%   9) 80%   15) 50%
4) 200%   10) 75%   16) 25%
5) 200%   11) 50%   17) 35%
6) 80%   12) 22%

Simple Interest

1) $158.40   7) $1,820.00   13) $3,936.00
2) $1,176.00   8) $161.50   14) 3 years
3) $1,500.00   9) $52.20   15) 5 years
4) $88.40   10) $945.00   16) 4 years
5) $28.00   11) $20,130.00
6) $9,072.00   12) $1,620.00

HiSET Subject Test – Mathematics

Chapter 4 : Exponents and Radicals Expressions

Topics that you'll practice in this chapter:

- ✓ Multiplication Property of Exponents
- ✓ Zero and Negative Exponents
- ✓ Division Property of Exponents
- ✓ Powers of Products and Quotients
- ✓ Negative Exponents and Negative Bases
- ✓ Scientific Notation
- ✓ Square Roots
- ✓ Simplifying Radical Expressions

Mathematics is no more computation than typing is literature.
– John Allen Paulos

HiSET Subject Test – Mathematics

Multiplication Property of Exponents

✎ Simplify and write the answer in exponential form.

1) $4 \times 4^5 =$

2) $8^4 \times 8 =$

3) $7^3 \times 7^3 =$

4) $9^2 \times 9^2 =$

5) $2^2 \times 2^4 \times 2 =$

6) $5 \times 5^3 \times 5^3 =$

7) $4^3 \times 4^2 \times 4 \times 4 =$

8) $5x \times x =$

9) $x^3 \times x^3 =$

10) $x^7 \times x^2 =$

11) $x^4 \times x^3 \times x^2 =$

12) $10x \times 3x =$

13) $4x^3 \times 4x^3 =$

14) $7x^3 \times x =$

15) $3x^2 \times 4x^2 \times x^2 =$

16) $5x^4 \times x^4 =$

17) $2x^8 \times 2x =$

18) $6x \times x^5 =$

19) $4x^2 \times 6x^6 =$

20) $5yx^3 \times 4x =$

21) $7x^3 \times y^5 x^7 =$

22) $y^2 x^3 \times y^5 x^4 =$

23) $3x^5 \times 4x^3 y^4 =$

24) $4x^4 \times 9x^2 y^5 =$

25) $5x^3 y^4 \times 6x^8 y^2 =$

26) $8x^3 y^6 \times 4xy^3 =$

27) $2xy^5 \times 6x^3 y^3 =$

28) $4x^5 y^2 \times 4x^2 y^8 =$

29) $7x \times 3y^8 x^2 \times y^5 =$

30) $x^3 \times 2y^3 x^4 \times 2y =$

31) $3yx^4 \times 3y^4 x \times 3xy^3 =$

32) $6y^3 \times 2y^2 x^4 \times 10yx^5 =$

WWW.MathNotion.Com

HiSET Subject Test – Mathematics

Zero and Negative Exponents

✎ **Evaluate the following expressions.**

1) $1^{-5} =$

2) $4^{-1} =$

3) $0^{10} =$

4) $1^{15} =$

5) $5^{-2} =$

6) $3^{-3} =$

7) $9^{-1} =$

8) $10^{-2} =$

9) $12^{-2} =$

10) $2^{-5} =$

11) $3^{-4} =$

12) $2^{-4} =$

13) $6^{-3} =$

14) $10^{-3} =$

15) $30^{-1} =$

16) $15^{-2} =$

17) $4^{-3} =$

18) $2^{-7} =$

19) $5^{-3} =$

20) $4^{-4} =$

21) $3^{-5} =$

22) $10^{-4} =$

23) $2^{-10} =$

24) $8^{-3} =$

25) $20^{-2} =$

26) $14^{-2} =$

27) $9^{-3} =$

28) $100^{-2} =$

29) $5^{-4} =$

30) $4^{-6} =$

31) $\left(\frac{1}{4}\right)^{-3} =$

32) $\left(\frac{1}{6}\right)^{-2} =$

33) $\left(\frac{1}{7}\right)^{-2} =$

34) $\left(\frac{2}{3}\right)^{-3} =$

35) $\left(\frac{1}{13}\right)^{-2} =$

36) $\left(\frac{7}{12}\right)^{-2} =$

37) $\left(\frac{1}{6}\right)^{-3} =$

38) $\left(\frac{1}{300}\right)^{-2} =$

39) $\left(\frac{2}{9}\right)^{-2} =$

40) $\left(\frac{7}{5}\right)^{-1} =$

41) $\left(\frac{13}{23}\right)^{0} =$

42) $\left(\frac{1}{4}\right)^{-5} =$

HiSET Subject Test – Mathematics

Division Property of Exponents

✎ **Simplify.**

1) $\dfrac{5^6}{5^7} =$

2) $\dfrac{8^8}{8^6} =$

3) $\dfrac{4^5}{4} =$

4) $\dfrac{3}{3^5} =$

5) $\dfrac{x}{x^6} =$

6) $\dfrac{3 \times 3^2}{3^2 \times 3^5} =$

7) $\dfrac{9^4}{9^2} =$

8) $\dfrac{10 \times 10^9}{10^2 \times 10^7} =$

9) $\dfrac{7^5 \times 7^7}{7^4 \times 7^8} =$

10) $\dfrac{15x}{30x^6} =$

11) $\dfrac{3x^9}{4x^4} =$

12) $\dfrac{15x^8}{10x^9} =$

13) $\dfrac{42x^5}{6y^9} =$

14) $\dfrac{36y^8}{4x^4y^5} =$

15) $\dfrac{2x^7}{9x} =$

16) $\dfrac{49x^8y^6}{7x^9} =$

17) $\dfrac{48x^2}{24x^6y^{12}} =$

18) $\dfrac{30yx^5}{6yx^7} =$

19) $\dfrac{19x^7y}{38x^{12}y^4} =$

20) $\dfrac{9x^8}{63x^8} =$

21) $\dfrac{9x^{-9}}{4x^{-3}} =$

WWW.MathNotion.Com

Powers of Products and Quotients

✎ **Simplify.**

1) $(4^3)^2 =$

2) $(2^3)^4 =$

3) $(2 \times 2^3)^2 =$

4) $(5 \times 5^5)^6 =$

5) $(19^4 \times 19^2)^3 =$

6) $(2^3 \times 2^4)^4 =$

7) $(5 \times 5^2)^2 =$

8) $(4^4)^4 =$

9) $(8x^5)^2 =$

10) $(3x^2y^4)^4 =$

11) $(7x^5y^2)^2 =$

12) $(5x^4y^4)^3 =$

13) $(2x^3y^3)^5 =$

14) $(10x^3y^4)^3 =$

15) $(13y^3y)^2 =$

16) $(5x^6x^4)^2 =$

17) $(6x^7y^6)^3 =$

18) $(12x^5x^7)^2 =$

19) $(2x^4 \times 2x)^4 =$

20) $(2x^4y^3)^5 =$

21) $(15x^7y^2)^2 =$

22) $(8x^3y^5)^3 =$

23) $(3x \times 2y^2)^4 =$

24) $\left(\dfrac{4x}{x^5}\right)^2 =$

25) $\left(\dfrac{x^4y^5}{x^3y^5}\right)^9 =$

26) $\left(\dfrac{36xy}{6x^5}\right)^3 =$

27) $\left(\dfrac{x^7}{x^8y^2}\right)^6 =$

28) $\left(\dfrac{xy^4}{x^3y^6}\right)^{-3} =$

29) $\left(\dfrac{5xy^8}{x^3}\right)^2 =$

30) $\left(\dfrac{xy^6}{2xy^3}\right)^{-4} =$

HiSET Subject Test – Mathematics

Negative Exponents and Negative Bases

✏️ **Simplify.**

1) $-9^{-1} =$

2) $-9^{-2} =$

3) $-2^{-5} =$

4) $-x^{-7} =$

5) $11x^{-1} =$

6) $-8x^{-3} =$

7) $-12x^{-5} =$

8) $-9x^{-8}y^{-6} =$

9) $32x^{-5}y^{-1} =$

10) $10a^{-9}b^{-3} =$

11) $-17x^4y^{-6} =$

12) $-\dfrac{25}{x^{-5}} =$

13) $-\dfrac{13x}{a^{-7}} =$

14) $\left(-\dfrac{1}{3}\right)^{-4} =$

15) $\left(-\dfrac{3}{4}\right)^{-2} =$

16) $-\dfrac{14}{a^{-6}b^{-3}} =$

17) $-\dfrac{7x}{x^{-8}} =$

18) $-\dfrac{a^{-9}}{b^{-5}} =$

19) $-\dfrac{11}{x^{-5}} =$

20) $\dfrac{8b}{-16c^{-6}} =$

21) $\dfrac{12ab}{a^{-4}b^{-3}} =$

22) $-\dfrac{8n^{-4}}{32p^{-7}} =$

23) $\dfrac{16ab^{-6}}{-6c^{-5}} =$

24) $\left(\dfrac{10a}{5c}\right)^{-4} =$

25) $\left(-\dfrac{12x}{4yz}\right)^{-3} =$

26) $\dfrac{8ab^{-7}}{-5c^{-3}} =$

27) $\left(-\dfrac{x^4}{x^5}\right)^{-5} =$

28) $\left(-\dfrac{x^{-2}}{7x^3}\right)^{-2} =$

29) $\left(-\dfrac{x^{-4}}{x^2}\right)^{-6} =$

HiSET Subject Test – Mathematics

Scientific Notation

✎ Write each number in scientific notation.

1) $0.223 =$

2) $0.09 =$

3) $4.5 =$

4) $900 =$

5) $2,000 =$

6) $0.006 =$

7) $33 =$

8) $9,400 =$

9) $1,470 =$

10) $52,000 =$

11) $8,000,000 =$

12) $0.00009 =$

13) $2,158,000 =$

14) $0.0039 =$

15) $0.000075 =$

16) $4,300,000 =$

17) $130,000 =$

18) $4,000,000,000 =$

19) $0.00009 =$

20) $0.0039 =$

✎ Write each number in standard notation.

21) $4 \times 10^{-1} =$

22) $1.2 \times 10^{-3} =$

23) $2.7 \times 10^{5} =$

24) $6 \times 10^{-4} =$

25) $3.6 \times 10^{-3} =$

26) $5.5 \times 10^{5} =$

27) $3.2 \times 10^{4} =$

28) $3.88 \times 10^{6} =$

29) $7 \times 10^{-6} =$

30) $4.2 \times 10^{-7} =$

WWW.MathNotion.Com

HiSET Subject Test – Mathematics

Square Roots

✎ **Find the value each square root.**

1) $\sqrt{16} =$ ___

2) $\sqrt{25} =$ ___

3) $\sqrt{1} =$ ___

4) $\sqrt{64} =$ ___

5) $\sqrt{0} =$ ___

6) $\sqrt{196} =$ ___

7) $\sqrt{4} =$ ___

8) $\sqrt{256} =$ ___

9) $\sqrt{36} =$ ___

10) $\sqrt{289} =$ ___

11) $\sqrt{169} =$ ___

12) $\sqrt{144} =$ ___

13) $\sqrt{100} =$ ___

14) $\sqrt{1,600} =$ ___

15) $\sqrt{2,500} =$ ___

16) $\sqrt{324} =$ ___

17) $\sqrt{529} =$ ___

18) $\sqrt{20} =$ ___

19) $\sqrt{625} =$ ___

20) $\sqrt{18} =$ ___

21) $\sqrt{50} =$ ___

22) $\sqrt{1,024} =$ ___

23) $\sqrt{160} =$ ___

24) $\sqrt{32} =$ ___

✎ **Evaluate.**

25) $\sqrt{4} \times \sqrt{25} =$ _____

26) $\sqrt{36} \times \sqrt{49} =$ _____

27) $\sqrt{6} \times \sqrt{6} =$ _____

28) $\sqrt{13} \times \sqrt{13} =$ _____

29) $2\sqrt{5} \times 3\sqrt{5} =$ _____

30) $\sqrt{12} \times \sqrt{3} =$ _____

31) $\sqrt{13} + \sqrt{13} =$ _____

32) $\sqrt{10} + 2\sqrt{10} =$ _____

33) $12\sqrt{7} - 10\sqrt{7} =$ _____

34) $4\sqrt{10} \times 2\sqrt{10} =$ _____

35) $5\sqrt{3} \times 8\sqrt{3} =$ _____

36) $6\sqrt{3} - \sqrt{12} =$ _____

WWW.MathNotion.Com

HiSET Subject Test – Mathematics

Simplifying Radical Expressions

✎ **Simplify.**

1) $\sqrt{13x^2} =$

2) $\sqrt{75x^2} =$

3) $\sqrt[3]{27a} =$

4) $\sqrt{64x^5} =$

5) $\sqrt{216a} =$

6) $\sqrt[3]{63w^3} =$

7) $\sqrt{192x} =$

8) $\sqrt{125v} =$

9) $\sqrt[3]{128x^2} =$

10) $\sqrt{100x^9} =$

11) $\sqrt{16x^4} =$

12) $\sqrt[3]{500a^5} =$

13) $\sqrt{242} =$

14) $\sqrt{392p^3} =$

15) $\sqrt{8m^6} =$

16) $\sqrt{198x^3y^3} =$

17) $\sqrt{121x^5y^5} =$

18) $\sqrt{16a^6b^3} =$

19) $\sqrt{90x^5y^7} =$

20) $\sqrt[3]{64y^2x^6} =$

21) $10\sqrt{16x^4} =$

22) $6\sqrt{81x^2} =$

23) $\sqrt[3]{56x^2y^6} =$

24) $\sqrt[3]{1,000x^5y^7} =$

25) $8\sqrt{50a} =$

26) $\sqrt[4]{625x^8y} =$

27) $\sqrt{24x^4y^5r^3} =$

28) $5\sqrt{36x^4y^5z^8} =$

29) $3\sqrt[3]{343x^9y^7} =$

30) $5\sqrt{81a^5b^2c^9} =$

31) $\sqrt[4]{625x^8y^{16}} =$

WWW.MathNotion.Com

HiSET Subject Test – Mathematics

Answers of Worksheets

Multiplication Property of Exponents

1) 4^6
2) 8^5
3) 7^6
4) 9^4
5) 2^7
6) 5^7
7) 4^7
8) $5x^2$
9) x^6
10) x^9
11) x^9
12) $30x^2$
13) $16x^6$
14) $7x^4$
15) $12x^6$
16) $5x^8$
17) $4x^9$
18) $6x^6$
19) $24x^8$
20) $20x^4y$
21) $7x^{10}y^5$
22) x^7y^7
23) $12x^8y^4$
24) $36x^6y^5$
25) $30x^{11}y^6$
26) $32x^4y^9$
27) $12x^4y^8$
28) $16x^7y^{10}$
29) $21x^3y^{13}$
30) $4x^7y^4$
31) $27x^6y^8$
32) $120x^9y^6$

Zero and Negative Exponents

1) 1
2) $\frac{1}{4}$
3) 0
4) 1
5) $\frac{1}{25}$
6) $\frac{1}{27}$
7) $\frac{1}{9}$
8) $\frac{1}{100}$
9) $\frac{1}{144}$
10) $\frac{1}{32}$
11) $\frac{1}{81}$
12) $\frac{1}{16}$
13) $\frac{1}{216}$
14) $\frac{1}{1,000}$
15) $\frac{1}{30}$
16) $\frac{1}{225}$
17) $\frac{1}{64}$
18) $\frac{1}{128}$
19) $\frac{1}{125}$
20) $\frac{1}{256}$
21) $\frac{1}{243}$
22) $\frac{1}{10,000}$
23) $\frac{1}{1,024}$
24) $\frac{1}{512}$
25) $\frac{1}{400}$
26) $\frac{1}{196}$
27) $\frac{1}{729}$
28) $\frac{1}{10,000}$
29) $\frac{1}{625}$
30) $\frac{1}{4,096}$
31) 64
32) 36
33) 49
34) $\frac{27}{8}$
35) 169
36) $\frac{144}{49}$
37) 216
38) $90,000$
39) $\frac{81}{4}$
40) $\frac{5}{7}$
41) 1
42) $1,024$

Division Property of Exponents

1) $\frac{1}{5}$
2) 8^2
3) 4^4
4) $\frac{1}{3^4}$
5) $\frac{1}{x^5}$
6) $\frac{1}{3^4}$
7) 9^2
8) 10
9) 1
10) $\frac{1}{2x^5}$
11) $\frac{3x^5}{4}$
12) $\frac{3}{2x}$
13) $\frac{7x^5}{y^9}$
14) $\frac{9y^3}{x^4}$

WWW.MathNotion.Com

HiSET Subject Test – Mathematics

15) $\frac{2x^6}{9}$

16) $\frac{7y^6}{x}$

17) $\frac{2}{x^4 y^{12}}$

18) $\frac{5}{x^2}$

19) $\frac{1}{2x^5 y^3}$

20) $\frac{1}{7}$

21) $\frac{9}{4x^6}$

Powers of Products and Quotients

1) 4^6

2) 2^{12}

3) 2^8

4) 5^{36}

5) 19^{18}

6) 2^{28}

7) 5^6

8) 4^{16}

9) $64x^{10}$

10) $81x^8 y^{16}$

11) $49x^{10} y^4$

12) $125x^{12} y^{12}$

13) $32x^{15} y^{15}$

14) $1,000x^9 y^{12}$

15) $169y^8$

16) $25x^{20}$

17) $216x^{21} y^{18}$

18) $144x^{24}$

19) $256x^{20}$

20) $32x^{20} y^{15}$

21) $225x^{14} y^4$

22) $512x^9 y^{15}$

23) $1,296x^4 y^8$

24) $\frac{16}{x^8}$

25) x^9

26) $\frac{216y^3}{x^{12}}$

27) $\frac{1}{x^6 y^{12}}$

28) $x^6 y^6$

29) $\frac{25y^{16}}{x^4}$

30) $\frac{16}{y^{12}}$

Negative Exponents and Negative Bases

1) $-\frac{1}{9}$

2) $-\frac{1}{81}$

3) $-\frac{1}{32}$

4) $-\frac{1}{x^7}$

5) $\frac{11}{x}$

6) $-\frac{8}{x^3}$

7) $-\frac{12}{x^5}$

8) $-\frac{9}{x^8 y^6}$

9) $\frac{32}{x^5 y}$

10) $\frac{10}{a^9 b^3}$

11) $-\frac{17x^4}{y^6}$

12) $-25x^5$

13) $-13xa^7$

14) 81

15) $\frac{16}{9}$

16) $-14a^6 b^3$

17) $-7x^9$

18) $-\frac{b^5}{a^9}$

19) $-11x^5$

20) $-\frac{bc^6}{2}$

21) $12a^5 b^4$

22) $-\frac{p^7}{4n^4}$

23) $-\frac{8ac^5}{3b^6}$

24) $\frac{c^4}{16a^4}$

25) $\frac{y^3 z^3}{27x^3}$

26) $-\frac{8ac^3}{5b^7}$

27) $-x^5$

28) $49x^{10}$

29) x^{36}

Scientific Notation

1) 2.23×10^{-1}

2) 9×10^{-2}

3) 4.5×10^0

4) 9×10^2

5) 2×10^3

6) 6×10^{-3}

7) 3.3×10^1

8) 9.4×10^3

9) 1.47×10^3

WWW.MathNotion.Com

HiSET Subject Test – Mathematics

10) 5.2×10^4
11) 8×10^6
12) 9×10^{-5}
13) 2.158×10^6
14) 3.9×10^{-3}
15) 7.5×10^{-5}
16) 4.3×10^6

17) 1.3×10^5
18) 4×10^9
19) 9×10^{-5}
20) 3.9×10^{-3}
21) 0.4
22) 0.0012
23) $270{,}000$

24) 0.0006
25) 0.0036
26) $550{,}000$
27) $32{,}000$
28) $3{,}880{,}000$
29) 0.000007
30) 0.00000042

Square Roots

1) 4
2) 5
3) 1
4) 8
5) 0
6) 14
7) 2
8) 16
9) 6

10) 17
11) 13
12) 12
13) 10
14) 40
15) 50
16) 18
17) 23
18) $2\sqrt{5}$

19) 25
20) $3\sqrt{2}$
21) $5\sqrt{2}$
22) 32
23) $4\sqrt{10}$
24) $4\sqrt{2}$
25) 10
26) 42
27) 6

28) 13
29) 30
30) 6
31) $2\sqrt{13}$
32) $3\sqrt{10}$
33) $2\sqrt{7}$
34) 80
35) 120
36) $4\sqrt{3}$

Simplifying radical expressions

1) $x\sqrt{13}$
2) $5x\sqrt{3}$
3) $3\sqrt[3]{a}$
4) $8x^2\sqrt{x}$
5) $6\sqrt{6a}$
6) $w\sqrt[3]{63}$
7) $8\sqrt{3x}$
8) $5\sqrt{5v}$
9) $4\sqrt[3]{2x^2}$
10) $10x^4\sqrt{x}$
11) $4x^2$

12) $5a\sqrt[3]{4a^2}$
13) $11\sqrt{2}$
14) $14p\sqrt{2p}$
15) $2m^3\sqrt{2}$
16) $3x.y\sqrt{22xy}$
17) $11x^2y^2\sqrt{xy}$
18) $4a^3b\sqrt{b}$
19) $3x^2y^3\sqrt{10xy}$
20) $4x^2\sqrt[3]{y^2}$
21) $40x^2$
22) $54x$

23) $2y^2\sqrt[3]{7x^2}$
24) $10xy^2\sqrt[3]{x^2y}$
25) $40\sqrt{2a}$
26) $5x^2\sqrt[4]{y}$
27) $2x^2y^2r\sqrt{6yr}$
28) $30x^2y^2z^4\sqrt{y}$
29) $21x^3y^2\sqrt[3]{y}$
30) $45a^2bc^4\sqrt{ac}$
31) $5x^2y^4$

HiSET Subject Test – Mathematics

Chapter 5:
Algebraic Expressions

Topics that you'll practice in this chapter:

- ✓ Simplifying Variable Expressions
- ✓ Simplifying Polynomial Expressions
- ✓ Translate Phrases into an Algebraic Statement
- ✓ The Distributive Property
- ✓ Evaluating One Variable Expressions
- ✓ Evaluating Two Variables Expressions
- ✓ Combining like Terms

Mathematics is, as it were, a sensuous logic, and relates to philosophy as do the arts, music, and plastic art to poetry. — K. Shegel

WWW.MathNotion.Com

HiSET Subject Test – Mathematics

Simplifying Variable Expressions

✎ **Simplify each expression.**

1) $3(x + 5) =$

2) $(-4)(7x - 5) =$

3) $11x + 5 - 6x =$

4) $-4 - 2x^2 - 6x^2 =$

5) $7 + 13x^2 + 3 =$

6) $3x^2 + 7x + 15x^2 =$

7) $3x^2 - 12x^2 + 4x =$

8) $4x^2 - 8x - 2x =$

9) $6x + 7(3 - 4x) =$

10) $8x + 4(15x - 3) =$

11) $6(-3x - 9) - 17 =$

12) $-11x^2 - (-5x) =$

13) $2x + 7 + 5 - 8x =$

14) $7 + 6x - 11 - 5x =$

15) $27x + 8 - 13 - 5x =$

16) $(-11)(-5x + 2) - 41x =$

17) $19x - 4(4 - 2x) =$

18) $16x + 3(3x + 6) + 10 =$

19) $5(-2x - 4) - 13x =$

20) $16x - 3x(x + 10) =$

21) $17x + 5x(2 - 4x) =$

22) $5x(-4x - 7) + 20x =$

23) $25x - 19 + 4x^2 =$

24) $6x(x - 11) + 25 =$

25) $4x - 5 + 15x + 3x^2 =$

26) $-7x^2 - 11x - 9x =$

27) $10x - 9x^2 - 3x^2 - 7 =$

28) $13 + 3x^2 - 9x^2 - 21x =$

29) $22x + 10x^2 - 15x + 17 =$

30) $4x^2 + 25x + 21x^2 =$

31) $29 - 12x^2 - 23x - 4x^2 =$

32) $22x - 19x - 9x^2 + 30 =$

WWW.MathNotion.Com

HiSET Subject Test – Mathematics

Simplifying Polynomial Expressions

✎ Simplify each polynomial.

1) $(2x^3 + 8x^2) - (11x + 3x^2) =$ _____

2) $(2x^5 + 7x^3) - (5x^3 + 11x^2) =$ _____

3) $(41x^4 + 5x^2) - (4x^2 + 20x^4) =$ _____

4) $13x - 8x^2 + 4(4x^2 + 3x^3) =$ _____

5) $(4x^3 - 22) + 5(3x^2 - 6x^3) =$ _____

6) $(4x^3 - 3x) - 5(2x^3 + x^4) =$ _____

7) $5(5x - 2x^3) - 2(8x^3 + 5x^2) =$ _____

8) $(3x^2 - 10x) - (5x^3 + 14x^2) =$ _____

9) $5x^3 - (3x^4 + 5x) + 2x^2 =$ _____

10) $11x^4 - (3x^2 + 5x) + 7x =$ _____

11) $(6x^2 - 3x^4) - (10x^4 + 3x^2) =$ _____

12) $2x^2 - 7x^3 + 19x^4 - 22x^3 =$ _____

13) $10x^2 - x^4 + 4x^4 - 32x^3 =$ _____

14) $-5x^2 + 17x^3 - 8x^2 - 6x =$ _____

15) $x^4 - 11x^5 - 30x^4 + 5x^2 =$ _____

16) $21x^3 + 13x - 5x^2 - 11x^3 =$ _____

WWW.MathNotion.Com

HiSET Subject Test – Mathematics

Translate Phrases into an Algebraic Statement

✍ **Write an algebraic expression for each phrase.**

1) 9 multiplied by x. _____

2) Subtract 11 from y. _____

3) 19 divided by x. _____

4) 38 decreased by y. _____

5) Add y to 40. _____

6) The square of 6. _____

7) x raised to the fifth power. _____

8) The sum of six and a number. _____

9) The difference between fifty–seven and y. _____

10) The quotient of nine and a number. _____

11) The quotient of the square of x and 25. _____

12) The difference between x and 6 is 19. _____

13) 10 times a reduced by the square of b. _____

14) Subtract the product of a and b from 41. _____

WWW.MathNotion.Com

HiSET Subject Test – Mathematics

The Distributive Property

✎ Use the distributive property to simply each expression.

1) $4(1 + 2x) =$

2) $2(4 + 7x) =$

3) $3(4x − 4) =$

4) $(2x − 5)(-6) =$

5) $(− 3)(x + 6) =$

6) $(4 + 3x)2 =$

7) $(−5)(8 − 3x) =$

8) $−(−5 − 7x) =$

9) $(−6x + 3)(−3) =$

10) $(−4)(x − 7) =$

11) $−(5 − 3x) =$

12) $3(9 + 4x) =$

13) $6(4 + 3x) =$

14) $(−5x + 3)2 =$

15) $(5 − 8x)(−3) =$

16) $(−12)(3x + 3) =$

17) $(5 − 3x)6 =$

18) $4(2 + 6x) =$

19) $8(7x − 3) =$

20) $(−2x + 3)4 =$

21) $(7 − 5x)(−9) =$

22) $(−10)(x − 8) =$

23) $(11 − 4x)3 =$

24) $(−6)(10x − 4) =$

25) $(3 − 9x)(−7) =$

26) $(−9)(x + 9) =$

27) $(−3 + 5x)(−7) =$

28) $(−5)(8 − 10x) =$

29) $12(4x − 8) =$

30) $(−10x + 13)(−3) =$

31) $(−8)(3x − 2) + 4(x + 5) =$

32) $(−8)(x + 4) − (6 + 5x) =$

WWW.MathNotion.Com

HiSET Subject Test – Mathematics

Evaluating One Variable Expressions

✍ Evaluate each expression using the value given.

1) $8 - x$, $x = 5$

2) $x - 9$, $x = 5$

3) $5x + 4$, $x = 3$

4) $x - 13$, $x = -4$

5) $12 - x$, $x = 4$

6) $x + 2$, $x = 6$

7) $4x + 8$, $x = 3$

8) $x + (-7)$, $x = -8$

9) $4x + 5$, $x = 2$

10) $3x + 9$, $x = -2$

11) $15 + 3x - 7$, $x = 2$

12) $17 - 3x$, $x = 3$

13) $8x - 9$, $x = 4$

14) $5x + 4$, $x = -3$

15) $10x + 5$, $x = 3$

16) $14 - 4x$, $x = -6$

17) $3(5x + 3)$, $x = 9$

18) $4(-3x - 6)$, $x = 3$

19) $7x - 2x + 12$, $x = 4$

20) $(5x + 6) \div 2$, $x = 8$

21) $(x + 18) \div 10$, $x = 12$

22) $5x - 12 + 3x$, $x = -3$

23) $(6 - 4x)(-3)$, $x = -4$

24) $9x^2 + 3x - 6$, $x = 2$

25) $x^2 - 10x$, $x = -5$

26) $3x(7 - 2x)$, $x = 2$

27) $12x + 6 - 2x^2$, $x = -4$

28) $(-3)(4x - 8 + 3x)$, $x = 3$

29) $(-6) + \frac{x}{4} + 3x$, $x = 16$

30) $(-6) + \frac{x}{5}$, $x = 35$

31) $\left(-\frac{45}{x}\right) - 7 + 2x$, $x = 9$

32) $\left(-\frac{21}{x}\right) - 12 + 4x$, $x = 7$

WWW.MathNotion.Com

HiSET Subject Test – Mathematics

Evaluating Two Variables Expressions

✏️ **Evaluate each expression using the values given.**

1) $2x - 4y$,

   $x = 4, y = 1$

2) $3x + 5y$,

   $x = -2, y = 2$

3) $-7a + 4b$,

   $a = 2, b = 4$

4) $3x + 5 - y$,

   $x = 5, y = 6$

5) $3z + 12 - 2k$,

   $z = 5, k = 6$

6) $6(-x - 3y)$,

   $x = 5, y = -2$

7) $5a + 3b$,

   $a = 3, b = 4$

8) $7x \div 3y$,

   $x = 3, y = 7$

9) $2x + 15 + 5y$,

   $x = -3, y = 1$

10) $5a - (18 - b)$,

    $a = 2, b = 8$

11) $2z + 20 + 5k$,

    $z = -6, k = 5$

12) $xy + 10 + 4x$,

    $x = 3, y = 5$

13) $2x + 4y - 8 + 5$,

    $x = 5, y = 2$

14) $\left(-\frac{24}{x}\right) + 3 + 2y$,

    $x = 4, y = 6$

15) $(-3)(-3a - 3b)$,

    $a = 4, b = 5$

16) $12 + 4x - 7 - y$,

    $x = 3, y = 5$

17) $11x + 5 - 8y + 6$,

    $x = 5, y = 2$

18) $10 + 2(-4x - 5y)$,

    $x = 5, y = 4$

19) $5x + 13 + 6y$,

    $x = 5, y = 6$

20) $10a - (7a + 3b) - 11$,

    $a = 3, b = 8$

Combining like Terms

✏️ **Simplify each expression.**

1) $11x + 3x + 6 =$

2) $8(2x - 6) =$

3) $18x - 7x + 11 =$

4) $(-4)(6x - 7) =$

5) $22x - 10x - 5 =$

6) $32x - 13 + 8x =$

7) $15 - (8x - 11) =$

8) $-24x + 17 - 11x =$

9) $12x - 8 - 6x + 9 =$

10) $21x + 5 - 36 + 12x =$

11) $28x + 3x - 11 =$

12) $(-3x + 4)5 =$

13) $2 + 4x + 9x - 8 =$

14) $6(2x - 5x) - 4 =$

15) $4(5x + 11) + 3x =$

16) $x - 14 - 11x =$

17) $5(10 + 9x) - 8x =$

18) $42x + 17 - 23x =$

19) $(-7x) + 19 + 20x =$

20) $(-7x) - 33 + 29x =$

21) $4(5x + 3) - 19x =$

22) $5(6 - 2x) - 15x =$

23) $-24x + (11 - 18x) =$

24) $(-9) - (6)(7x + 3) =$

25) $(-1)(8x - 10) - 21x =$

26) $-36x + 14 + 27x - 5x =$

27) $3(-13x + 6) - 17x =$

28) $-5x - 42 + 32x =$

29) $37x - 19x + 15 - 9x =$

30) $3(5x + 7x) - 31 =$

31) $14 - 6x - 15 - 9x =$

32) $-2(-5x - 7x) + 27x =$

HiSET Subject Test – Mathematics

Answers of Worksheets

Simplifying Variable Expressions

1) $3x + 15$
2) $-28x + 20$
3) $5x + 5$
4) $-8x^2 - 4$
5) $13x^2 + 10$
6) $18x^2 + 7x$
7) $-9x^2 + 4x$
8) $4x^2 - 10x$
9) $-22x + 21$
10) $68x - 12$
11) $-18x - 71$
12) $-11x^2 + 5x$
13) $-6x + 12$
14) $x - 4$
15) $22x - 5$
16) $14x - 22$
17) $27x - 16$
18) $25x + 28$
19) $-23x - 20$
20) $-3x^2 - 14x$
21) $-20x^2 + 27x$
22) $-20x^2 - 15x$
23) $4x^2 + 25x - 19$
24) $6x^2 - 66x + 25$
25) $3x^2 + 19x - 5$
26) $-7x^2 - 20x$
27) $-12x^2 + 10x - 7$
28) $-6x^2 - 21x + 13$
29) $10x^2 + 7x + 17$
30) $25x^2 + 25x$
31) $-16x^2 - 23x + 29$
32) $-9x^2 + 3x + 30$

Simplifying Polynomial Expressions

1) $2x^3 + 5x^2 - 11x$
2) $2x^5 + 2x^3 - 11x^2$
3) $21x^4 + x^2$
4) $12x^3 + 8x^2 + 13x$
5) $-26x^3 + 15x^2 - 22$
6) $-5x^4 - 6x^3 - 3x$
7) $-26x^3 - 10x^2 + 25x$
8) $-5x^3 - 11x^2 - 10x$
9) $-3x^4 + 5x^3 + 2x^2 - 5x$
10) $11x^4 - 3x^2 + 2x$
11) $-13x^4 + 3x^2$
12) $19x^4 - 29x^3 + 2x^2$
13) $3x^4 - 32x^3 + 10x^2$
14) $17x^3 - 13x^2 - 6x$
15) $-11x^5 - 29x^4 + 5x^2$
16) $10x^3 - 5x^2 + 13x$

Translate Phrases into an Algebraic Statement

1) $9x$
2) $y - 11$
3) $\frac{19}{x}$
4) $38 - y$
5) $y + 40$
6) 6^2
7) x^5
8) $6 + x$
9) $57 - y$
10) $\frac{9}{x}$
11) $\frac{x^2}{25}$
12) $x - 6 = 19$
13) $10a - b^2$
14) $41 - ab$

The Distributive Property

1) $8x + 4$
2) $14x + 8$
3) $12x - 12$
4) $-12x + 30$
5) $-3x - 18$
6) $6x + 8$
7) $15x - 40$
8) $7x + 5$
9) $18x - 9$
10) $-4x + 28$
11) $3x - 5$
12) $12x + 27$

HiSET Subject Test – Mathematics

13) $18x + 24$
14) $-10x + 6$
15) $24x - 15$
16) $-36x - 36$
17) $-18x + 30$

18) $24x + 8$
19) $56x - 24$
20) $-8x + 12$
21) $45x - 63$
22) $-10x + 80$

23) $-12x + 33$
24) $-60x + 24$
25) $63x - 21$
26) $-9x - 81$
27) $-35x + 21$

28) $50x - 40$
29) $48x - 96$
30) $30x - 39$
31) $-20x + 36$
32) $-13x - 38$

Evaluating One Variables

1) 3
2) -4
3) 19
4) -17
5) 8
6) 8
7) 20
8) -15

9) 13
10) 3
11) 14
12) 8
13) 23
14) -11
15) 35
16) 38

17) 144
18) -60
19) 32
20) 23
21) 3
22) -36
23) -66
24) 36

25) 75
26) 18
27) -74
28) -39
29) 46
30) 1
31) 6
32) 13

Evaluating Two Variables

1) 4
2) 4
3) 2
4) 14
5) 15

6) 6
7) 27
8) 1
9) 14
10) 0

11) 33
12) 37
13) 15
14) 9
15) 81

16) 12
17) 50
18) -70
19) 74
20) -26

Combining like Terms

1) $14x + 6$
2) $16x - 48$
3) $11x + 11$
4) $-24x + 28$
5) $12x - 5$
6) $40x - 13$
7) $-8x + 26$
8) $-35x + 17$

9) $6x + 1$
10) $33x - 31$
11) $31x - 11$
12) $-15x + 20$
13) $13x - 6$
14) $-18x - 4$
15) $23x + 44$
16) $-10x - 14$

17) $37x + 50$
18) $19x + 17$
19) $13x + 19$
20) $22x - 33$
21) $x + 12$
22) $-25x + 30$
23) $-42x + 11$
24) $-42x - 27$

25) $-29x + 10$
26) $-14x + 14$
27) $-56x + 18$
28) $27x - 42$
29) $9x + 15$
30) $36x - 31$
31) $-15x - 1$
32) $51x$

HiSET Subject Test – Mathematics

Chapter 6 :
Equations and Inequalities

Topics that you'll practice in this chapter:

- ✓ One–Step Equations
- ✓ Multi–Step Equations
- ✓ Graphing Single–Variable Inequalities
- ✓ One–Step Inequalities
- ✓ Multi-Step Inequalities
- ✓ Systems of Equations
- ✓ Systems of Equations Word Problems

"Life is a math equation. In order to gain the most, you have to know how to convert negatives into positives." – Anonymous

HiSET Subject Test – Mathematics

One–Step Equations

✎ Find the answer for each equation.

1) $3x = 90, x = $ ____

2) $5x = 35, x = $ ____

3) $6x = 24, x = $ ____

4) $24x = 144, x = $ ____

5) $x + 15 = 20, x = $ ____

6) $x - 7 = 4, x = $ ____

7) $x - 9 = 2, x = $ ____

8) $x + 15 = 23, x = $ ____

9) $x - 4 = 13, x = $ ____

10) $12 = 16 + x, x = $ ____

11) $x - 10 = 2, x = $ ____

12) $5 - x = -11, x = $ ____

13) $28 = -6 + x, x = $ ____

14) $x - 20 = -35, x = $ ____

15) $x + 14 = -4, x = $ ____

16) $14 = 28 - x, x = $ ____

17) $7 + x = -7, x = $ ____

18) $x - 16 = 4, x = $ ____

19) $30 = x - 15, x = $ ____

20) $x - 5 = -18, x = $ ____

21) $x - 10 = 24, x = $ ____

22) $x - 20 = -25, x = $ ____

23) $x - 17 = 30, x = $ ____

24) $-70 = x - 28, x = $ ____

25) $x - 9 = 13, x = $ ____

26) $36 = 4x, x = $ ____

27) $x - 35 = 25, x = $ ____

28) $x - 25 = 10, x = $ ____

29) $70 - x = 16, x = $ ____

30) $x - 10 = 14, x = $ ____

31) $17 - x = -13, x = $ ____

32) $x - 9 = -30, x = $ ____

WWW.MathNotion.Com

HiSET Subject Test – Mathematics

Multi–Step Equations

✏ **Find the answer for each equation.**

1) $3x + 3 = 9$

2) $-x + 5 = 12$

3) $4x - 8 = 8$

4) $-(3 - x) = 5$

5) $4x - 8 = 16$

6) $12x - 15 = 9$

7) $2x - 18 = 2$

8) $4x + 8 = 16$

9) $24x + 27 = 75$

10) $-14(3 + x) = 14$

11) $-3(2 + x) = 6$

12) $12 = -(x - 7)$

13) $3(3 - x) = 30$

14) $-15 = -(3x + 6)$

15) $40(3 + x) = 40$

16) $5(x - 10) = 25$

17) $-18 = x + 8x$

18) $3x + 25 = -2x - 10$

19) $7(6 + 3x) = -63$

20) $18 - 3x = -4 - 5x$

21) $4 - 6x = 36 + 2x$

22) $15 + 15x = -5 + 5x$

23) $42 = (-6x) - 7 + 7$

24) $21 = 3x - 21 + 4x$

25) $-18 = -6x - 9 + 3x$

26) $5x - 15 = -29 + 6x$

27) $7x - 18 = 4x + 3$

28) $-7 - 4x = 5(4 - x)$

29) $x - 5 = -5(-3 - x)$

30) $13x - 68 = 15x - 102$

31) $-5x - 3 = -3(9 + 3x)$

32) $-2x - 15 = 6x + 17$

WWW.MathNotion.Com

HiSET Subject Test – Mathematics

Graphing Single–Variable Inequalities

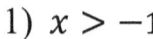 **Draw a graph for each inequality.**

1) $x > -1$

2) $x \leq 2$

3) $x \geq 0$

4) $x < -3$

5) $x < \frac{1}{2}$

6) $x \leq -2$

7) $x \leq 3$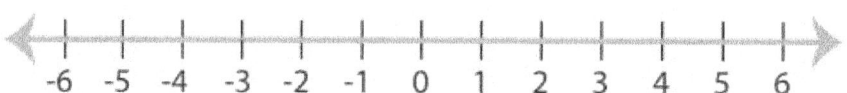

8) $x \geq -\frac{7}{2}$

WWW.MathNotion.Com

HiSET Subject Test – Mathematics

One–Step Inequalities

✏️ **Find the answer for each inequality and graph it.**

1) $x + 4 \geq 4$

2) $x - 5 \leq 2$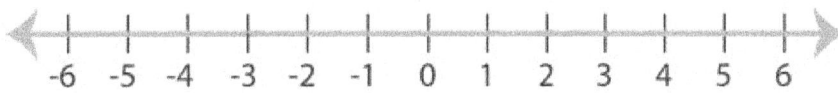

3) $5x > 35$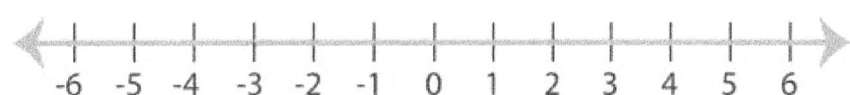

4) $9 + x \leq 11$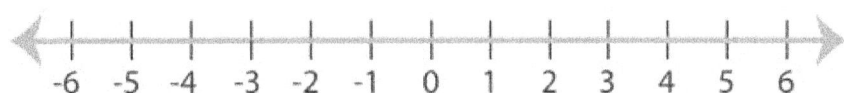

5) $x - 5 < -9$

6) $9x \geq 72$

7) $9x \leq 27$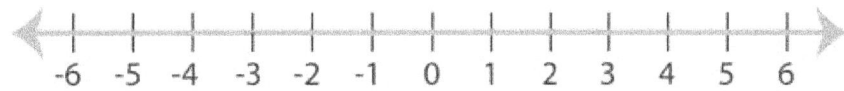

8) $x + 19 > 16$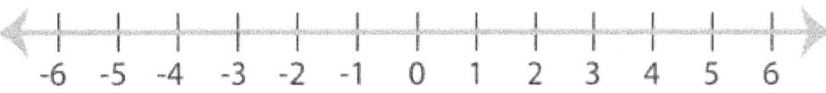

HiSET Subject Test – Mathematics

Multi-Step Inequalities

✎ Calculate each inequality.

1) $x - 3 \leq 7$

2) $8 - x \leq 8$

3) $3x - 9 \leq 9$

4) $4x - 4 \geq 8$

5) $x - 7 \geq 1$

6) $5x - 15 \leq 5$

7) $6x - 8 \leq 4$

8) $-11 + 6x \leq 12$

9) $4(x - 4) \leq 16$

10) $3x - 10 \leq 11$

11) $5x - 25 < 25$

12) $9x - 5 < 22$

13) $20 - 7x \geq -15$

14) $33 + 6x < 45$

15) $8 + 8x \geq 96$

16) $7 + 3x < 13$

17) $4x - 3 < 9$

18) $5(2 - 2x) \geq -30$

19) $-(7 + 6x) < 29$

20) $12 - 8x \geq -20$

21) $-4(x - 6) > 24$

22) $\dfrac{3x + 9}{6} \leq 10$

23) $\dfrac{4x - 10}{3} \leq 2$

24) $\dfrac{2x - 8}{3} > 2$

25) $8 + \dfrac{x}{6} < 9$

26) $\dfrac{9x}{7} - 4 < 5$

27) $\dfrac{15x + 45}{15} > 1$

28) $16 + \dfrac{x}{4} < 6$

WWW.MathNotion.Com

HiSET Subject Test – Mathematics

Systems of Equations

✎ **Calculate each system of equations.**

1) $-x + y = 2$ $x =$ ___
   $-4x + 2y = 6$ $y =$ ___

2) $-15x + 3y = -9$ $x =$ ___
   $9x - 16y = 48$ $y =$ ___

3) $y = -7$ $x =$ ___
   $6x + 5y = 7$ $y =$ ___

4) $3y = -9x + 15$ $x =$ ___
   $5x - 4y = -3$ $y =$ ___

5) $10x - 9y = -13$ $x =$ ___
   $-5x + 3y = 11$ $y =$ ___

6) $-12x - 16y = 20$ $x =$ ___
   $6x - 12y = 30$ $y =$ ___

7) $5x - 14y = -23$ $x =$ ___
   $-18x + 21y = 24$ $y =$ ___

8) $15x - 21y = -6$ $x =$ ___
   $2x - 3y = -2$ $y =$ ___

9) $-x + 3y = 3$ $x =$ ___
   $-14x + 16y = -10$ $y =$ ___

10) $x + 5y = 50$ $x =$ ___
    $3x + 10y = 80$ $y =$ ___

11) $6x - 7y = -8$ $x =$ ___
    $-x - 4y = -9$ $y =$ ___

12) $2x + 4y = -10$ $x =$ ___
    $2x - 8y = 14$ $y =$ ___

13) $4x + 3y = 12$ $x =$ ___
    $5x - 3y = 15$ $y =$ ___

14) $3x - 2y = 3$ $x =$ ___
    $7x - 8y = 22$ $y =$ ___

15) $3x + 2y = 5$ $x =$ ___
    $-10x - 4y = -14$ $y =$ ___

16) $10x + 7y = 1$ $x =$ ___
    $-5x - 7y = 24$ $y =$ ___

WWW.MathNotion.Com

HiSET Subject Test – Mathematics

Systems of Equations Word Problems

✍ **Find the answer for each word problem.**

1) Tickets to a movie cost $4 for adults and $3 for students. A group of friends purchased 8 tickets for $31.00. How many adults ticket did they buy? ____

2) At a store, Eva bought two shirts and five hats for $77.00. Nicole bought three same shirts and four same hats for $84.00. What is the price of each shirt? ____

3) A farmhouse shelters 18 animals, some are pigs, and some are ducks. Altogether there are 66 legs. How many pigs are there? ____

4) A class of 214 students went on a field trip. They took 36 vehicles, some cars and some buses. If each car holds 5 students and each bus hold 22 students, how many buses did they take? ____

5) A theater is selling tickets for a performance. Mr. Smith purchased 5 senior tickets and 3 child tickets for $105 for his friends and family. Mr. Jackson purchased 3 senior tickets and 5 child tickets for $79. What is the price of a senior ticket? $____

6) The difference of two numbers is 10. Their sum is 20. What is the bigger number? $____

7) The sum of the digits of a certain two-digit number is 7. Reversing its digits increase the number by 9. What is the number? ____

8) The difference of two numbers is 11. Their sum is 25. What are the numbers? _____

9) The length of a rectangle is 5 meters greater than 2 times the width. The perimeter of rectangle is 28 meters. What is the length of the rectangle? _____

10) Jim has 25 nickels and dimes totaling $1.80. How many nickels does he have? ____

WWW.MathNotion.Com

HiSET Subject Test – Mathematics

Answers of Worksheets

One–Step Equations

1) 30	9) 17	17) −14	25) 22
2) 7	10) −4	18) 20	26) 9
3) 4	11) 12	19) 45	27) 60
4) 6	12) 16	20) −13	28) 35
5) 5	13) 34	21) 34	29) 54
6) 11	14) −15	22) −5	30) 24
7) 11	15) −18	23) 47	31) 30
8) 8	16) 14	24) −42	32) −21

Multi–Step Equations

1) 2	9) 2	17) −2	25) 3
2) −7	10) −4	18) −7	26) 14
3) 4	11) −4	19) −5	27) 7
4) 8	12) −5	20) −11	28) 27
5) 6	13) −7	21) −4	29) −5
6) 2	14) 3	22) −2	30) 17
7) 10	15) −2	23) −7	31) −6
8) 2	16) 15	24) 6	32) −4

Graphing Single–Variable Inequalities

1)
2)
3)
4)

WWW.MathNotion.Com

HiSET Subject Test – Mathematics

5) [number line with open circle at 1/2, shaded right]

6) [number line with closed circle at -2, shaded right]

7) [number line with closed circle at 3, shaded right]

8) [number line with closed circle at -3.5, shaded left]

One–Step Inequalities

1) [number line with closed circle at 0, shaded right]

2) [number line with closed circle at 7, shaded left]

3) [number line with open circle at 7, shaded right]

4) [number line with closed circle at 2, shaded right]

5) [number line with open circle at -4, shaded left]

6) [number line with closed circle at 8, shaded right]

7) [number line with closed circle at 3, shaded left]

8) [number line with open circle at -3, shaded left]

Multi-Step Inequalities

1) $x \leq 10$
2) $x \geq 0$
3) $x \leq 6$
4) $x \geq 3$
5) $x \geq 8$
6) $x \leq 4$
7) $x \leq 2$
8) $x \leq \frac{23}{6}$
9) $x \leq 8$
10) $x \leq 7$
11) $x < 10$
12) $x < 3$
13) $x \leq 5$
14) $x < 2$
15) $x \geq 11$
16) $x < 2$
17) $x < 3$
18) $x \leq 4$
19) $x > -6$
20) $x \leq 4$
21) $x < 0$
22) $x \leq 17$
23) $x \leq 4$

WWW.MathNotion.Com

HiSET Subject Test – Mathematics

24) $x > 7$ 26) $x < 7$ 28) $x < -40$

25) $x < 6$ 27) $x > -2$

Systems of Equations

1) $x = -1, y = 1$ 7) $x = 1, y = 2$ 13) $x = 3, y = 0$

2) $x = 0, y = -3$ 8) $x = 8, y = 6$ 14) $x = -2, y = -\frac{9}{2}$

3) $x = 7$ 9) $x = 3, y = 2$ 15) $x = 1, y = 1$

4) $x = 1, y = 2$ 10) $x = -20, y = 14$ 16) $x = 5, y = -7$

5) $x = -4, y = -3$ 11) $x = 1, y = 2$

6) $x = 1, y = -2$ 12) $x = -1, y = -2$

Systems of Equations Word Problems

1) 7   5) $18   9) 11 meters

2) $16   6) 15   10) 14

3) 15   7) 34

4) 2   8) 18, 7

HiSET Subject Test – Mathematics

HiSET Subject Test – Mathematics

Chapter 7:

Linear Functions

Topics that you'll practice in this chapter:

- ✓ Finding Slope
- ✓ Graphing Lines Using Line Equation
- ✓ Writing Linear Equations
- ✓ Graphing Linear Inequalities
- ✓ Finding Midpoint
- ✓ Finding Distance of Two Points

"Nature is written in mathematical language." – Galileo Galilei

HiSET Subject Test – Mathematics

Finding Slope

✍ **Find the slope of each line.**

1) $y = x + 8$

2) $y = -3x + 5$

3) $y = 2x + 12$

4) $y = -4x + 19$

5) $y = 11 + 6x$

6) $y = 7 - 5x$

7) $y = 8x + 19$

8) $y = -9x + 20$

9) $y = -7x + 4$

10) $y = 3x - 8$

11) $y = \frac{1}{3}x + 8$

12) $y = -\frac{4}{5}x + 9$

13) $-3x + 6y = 30$

14) $4x + 4y = 16$

15) $3y - x = 10$

16) $8y - x = 5$

✍ **Find the slope of the line through each pair of points.**

17) $(2, 3), (7, 10)$

18) $(-3, 5), (2, 15)$

19) $(5, -3), (1, 9)$

20) $(-5, -5), (10, 25)$

21) $(22, 3), (7, 18)$

22) $(-16, 8), (-7, 26)$

23) $(25, 11), (29, 19)$

24) $(26, -19), (14, 17)$

25) $(22, -13), (20, -11)$

26) $(19, 7), (15, -3)$

27) $(5, 7), (11, 19)$

28) $(52, -62), (40, 70)$

WWW.MathNotion.Com

HiSET Subject Test – Mathematics

Graphing Lines Using Line Equation

✏️ **Sketch the graph of each line.**

1) $y = x - 2$

2) $y = -3x + 2$

3) $x + y = 0$

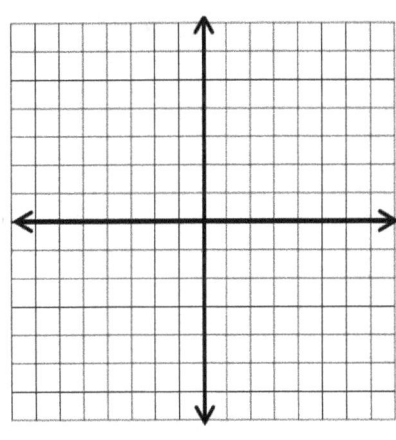
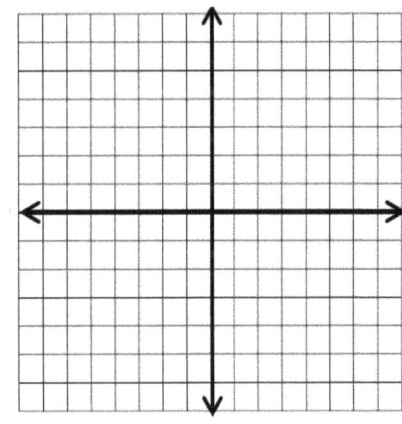
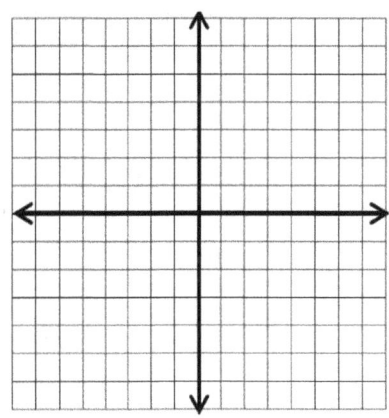

4) $x + y = -3$

5) $2x + 3y = -4$

6) $y - 3x + 6 = 0$

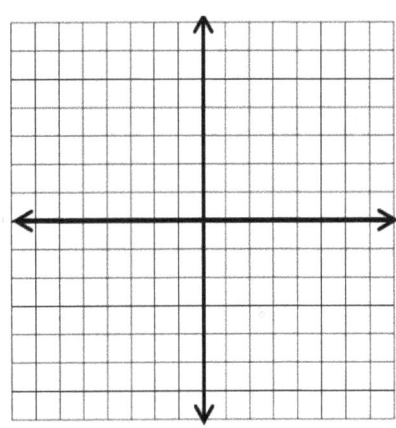
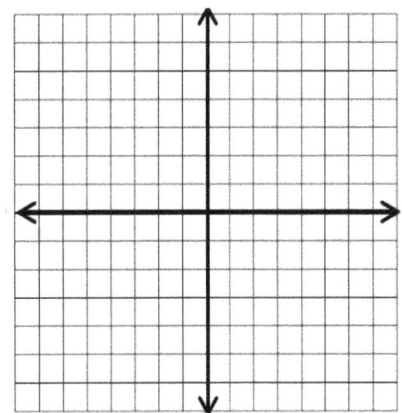
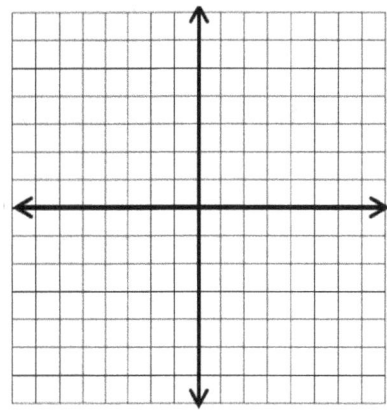

WWW.MathNotion.Com

HiSET Subject Test – Mathematics

Writing Linear Equations

✎ **Write the equation of the line through the given points.**

1) Through: $(2, -5), (3, 9)$

2) Through: $(-6, 3), (3, 12)$

3) Through: $(10, 7), (5, 27)$

4) Through: $(15, 11), (3, -1)$

5) Through: $(24, 17), (12, -7)$

6) Through: $(8, 29), (4, -7)$

7) Through: $(20, -16), (12, 0)$

8) Through: $(-3, 10), (2, -5)$

9) Through: $(-6, 17), (4, -3)$

10) Through: $(-8, 22), (5, -4)$

11) Through: $(9, 27), (3, -3)$

12) Through: $(11, 32), (9, 4)$

13) Through: $(-3, 13), (-4, 0)$

14) Through: $(-5, 5), (5, 15)$

15) Through: $(18, -32), (11, 3)$

16) Through: $(-4, 25), (4, -15)$

✎ **Find the answer for each problem.**

17) What is the equation of a line with slope 6 and intercept 12? _____

18) What is the equation of a line with slope −11 and intercept −4? _____

19) What is the equation of a line with slope −3 and passes through point $(5, 2)$? _____

20) What is the equation of a line with slope −5 and passes through point $(-2, -1)$? _____

21) The slope of a line is −10 and it passes through point $(-3, 0)$. What is the equation of the line? _____

22) The slope of a line is 8 and it passes through point $(0, 7)$. What is the equation of the line? _____

WWW.MathNotion.Com

HiSET Subject Test – Mathematics

Graphing Linear Inequalities

✎ **Sketch the graph of each linear inequality.**

1) $y > 4x - 5$

2) $y < 2x + 4$

3) $y \leq -5x - 2$

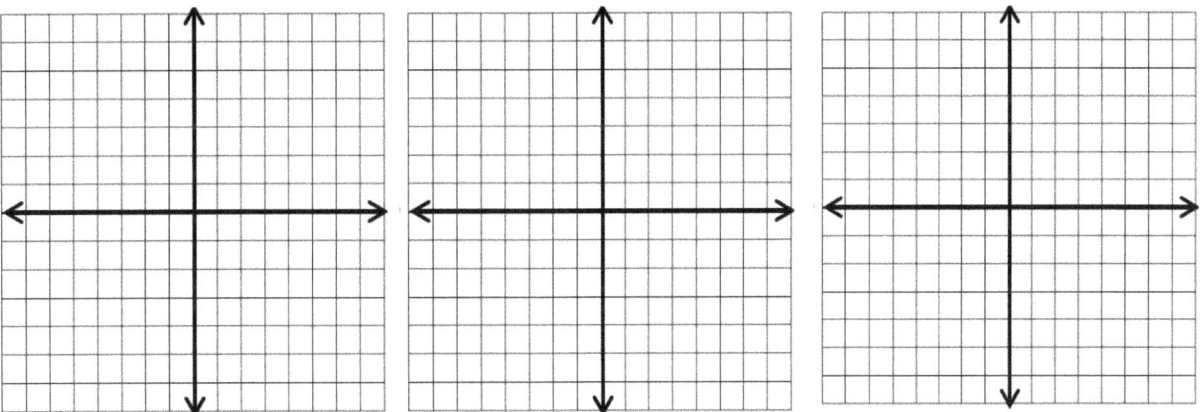

4) $4y \geq 12 + 4x$

5) $-12y < 3x - 24$

6) $5y \geq -15x + 10$

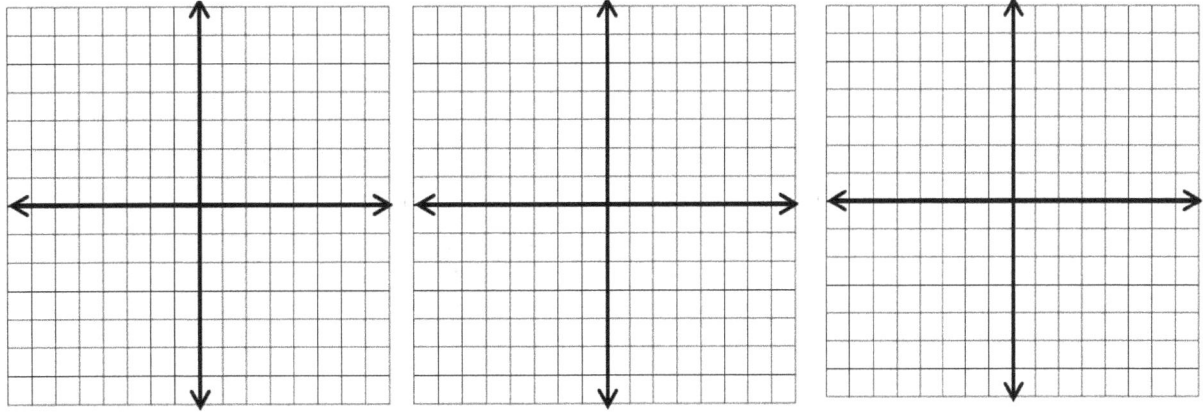

WWW.MathNotion.Com

HiSET Subject Test – Mathematics

Finding Midpoint

✎ **Find the midpoint of the line segment with the given endpoints.**

1) $(-4, -3), (2, 3)$

2) $(9, 0), (-1, 8)$

3) $(9, -6), (3, 14)$

4) $(-10, -6), (0, 8)$

5) $(2, -5), (14, -15)$

6) $(-10, -3), (4, -13)$

7) $(8, 7), (-8, 13)$

8) $(-3, 6), (-9, 2)$

9) $(-4, 5), (16, -9)$

10) $(7, 14), (9, -2)$

11) $(-8, 6), (6, 6)$

12) $(10, 5), (-2, -3)$

13) $(-5, 12), (-3, 3)$

14) $(12, 7), (8, -2)$

15) $(10, 2), (-6, 14)$

16) $(-1, -2), (-7, 10)$

17) $(7, -7), (13, -13)$

18) $(-3, -8), (11, -4)$

19) $(5, -11), (-8, 9)$

20) $(14, -4), (16, 14)$

21) $(0, -5), (8, -1)$

22) $(3, 0), (-21, 18)$

23) $(17, -3), (-7, -5)$

24) $(26, -12), (6, 24)$

✎ **Find the answer for each problem.**

25) One endpoint of a line segment is $(-3, 7)$ and the midpoint of the line segment is $(-6, 9)$. What is the other endpoint? _____

26) One endpoint of a line segment is $(-3, 7)$ and the midpoint of the line segment is $(1, 5)$. What is the other endpoint? _____

27) One endpoint of a line segment is $(-10, -16)$ and the midpoint of the line segment is $(2, 9)$. What is the other endpoint? _____

HiSET Subject Test – Mathematics

Finding Distance of Two Points

✎ Find the distance between each pair of points.

1) (6, 3), (−3, −9)

2) (5, 2), (−10, −6)

3) (8, 5), (8, 3)

4) (−8, −2), (2, 22)

5) (6, −7), (−3, −7)

6) (12, 0), (−9, −20)

7) (3, 20), (3, −5)

8) (10, 17), (5, 5)

9) (7, −2), (−4, −2)

10) (13, 4), (5, −2)

11) (11, 13), (5, 5)

12) (1, 4), (−23, −3)

13) (9, 8), (5, −4)

14) (−11, −4), (5, 8)

15) (−2, −6), (−2, −12)

16) (−1, −4), (23, 3)

17) (19, 3), (7, −6)

18) (−5, −2), (3, 4)

19) (2, 6), (2, −12)

20) (−4, −2), (8, −2)

✎ Find the answer for each problem.

21) Triangle ABC is a right triangle on the coordinate system and its vertices are (−2, 5), (−2, 1), and (1, 1). What is the area of triangle ABC? _____

22) Three vertices of a triangle on a coordinate system are (3, −6), (−5, −12), and (3, −18). What is the perimeter of the triangle? _____

23) Four vertices of a rectangle on a coordinate system are (−2, 2), (−2, 6), (4, 2), and (4, 6). What is its perimeter? _____

WWW.MathNotion.Com

HiSET Subject Test – Mathematics

Answers of Worksheets

Finding Slope

1) 1
2) -3
3) 2
4) -4
5) 6
6) -5
7) 8
8) -9

9) -7
10) 3
11) $\frac{1}{3}$
12) $-\frac{4}{5}$
13) $\frac{1}{2}$
14) -1
15) $\frac{1}{3}$

16) $\frac{1}{8}$
17) $\frac{7}{5}$
18) 2
19) -3
20) 2
21) -1
22) 2

23) 2
24) -3
25) -1
26) $\frac{5}{2}$
27) 2
28) -11

Graphing Lines Using Line Equation

1) $y = x - 2$

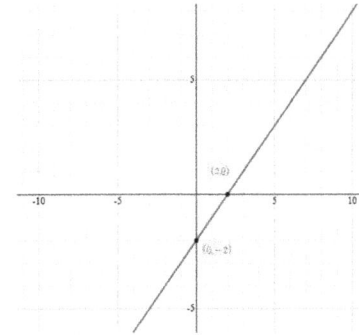

2) $y = -3x + 2$

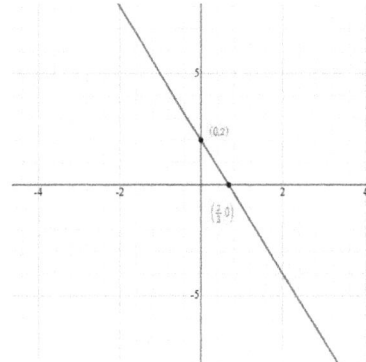

3) $x + y = 0$

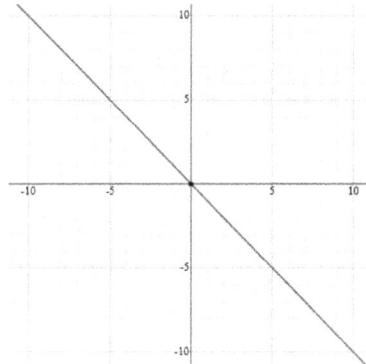

4) $x + y = -3$

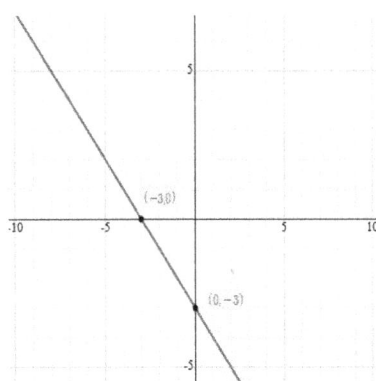

5) $2x + 3y = -4$

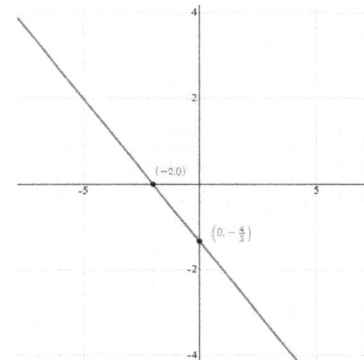

6) $y - 3x + 6 = 0$

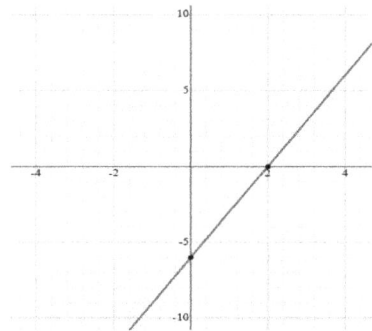

WWW.MathNotion.Com

HiSET Subject Test – Mathematics

Writing Linear Equations

1) $y = 14x - 33$
2) $y = x + 9$
3) $y = -4x + 47$
4) $y = x - 4$
5) $y = 2x - 31$
6) $y = 9x - 43$
7) $y = -2x + 24$
8) $y = -3x + 1$
9) $y = -2x + 5$
10) $y = -2x + 6$
11) $y = 5x - 18$
12) $y = 14x - 122$
13) $y = 13x + 52$
14) $y = x + 10$
15) $y = -5x + 58$
16) $y = -5x + 5$
17) $y = 6x + 12$
18) $y = -11x - 4$
19) $y = -3x + 17$
20) $y = -5x - 11$
21) $y = -10x - 30$
22) $y = 8x + 7$

Graphing Linear Inequalities

1) $y > 4x - 5$

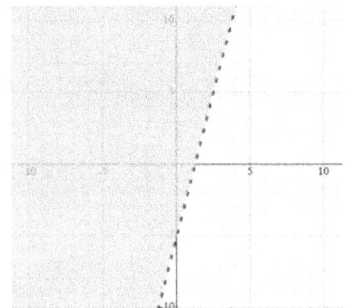

2) $y < 2x + 4$

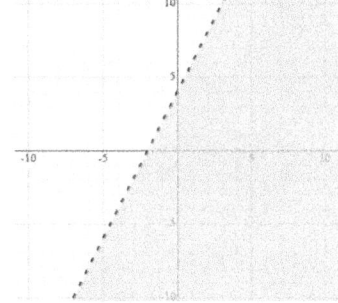

3) $y \leq -5x - 2$

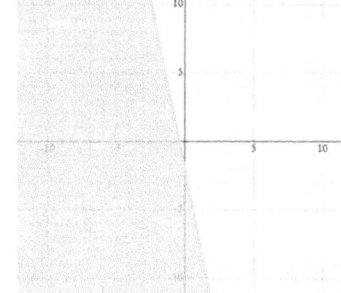

4) $4y \geq 12 + 4x$

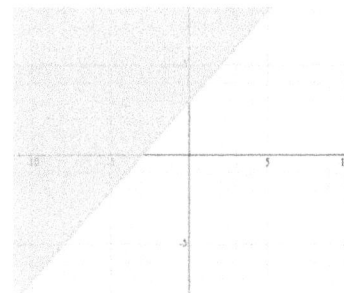

5) $-12y < 3x - 24$

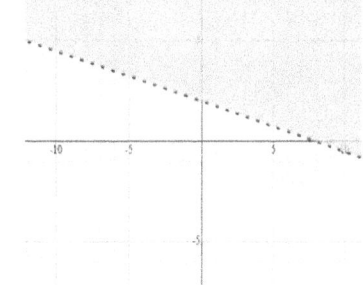

6) $5y \geq -15x + 10$

Finding Midpoint

1) $(-1, 0)$
2) $(4, 4)$
3) $(6, 4)$
4) $(-5, 1)$
5) $(8, -10)$
6) $(-3, -8)$
7) $(0, 10)$
8) $(-6, 4)$
9) $(6, -2)$
10) $(8, 6)$
11) $(-1, 6)$
12) $(4, 1)$
13) $(-4, 7.5)$
14) $(10, 2.5)$
15) $(2, 8)$
16) $(-4, 4)$
17) $(10, -10)$
18) $(4, -6)$

WWW.MathNotion.Com   95

HiSET Subject Test – Mathematics

19) $(-1.5, -1)$
20) $(15, 5)$
21) $(4, -3)$

22) $(-9, 9)$
23) $(5, -4)$
24) $(16, 6)$

25) $(-9, 11)$
26) $(5, 3)$
27) $(14, 34)$

Finding Distance of Two Points

1) 15
2) 17
3) 2
4) 26
5) 9
6) 29
7) 25
8) 13

9) 11
10) 10
11) 10
12) 25
13) $4\sqrt{10}$
14) 20
15) 6
16) 25

17) 15
18) 10
19) 18
20) 12
21) 6 *square units*
22) 32 *units*
23) 20 *units*

Chapter 8: Polynomials

Topics that you'll practice in this chapter:

- ✓ Writing Polynomials in Standard Form
- ✓ Simplifying Polynomials
- ✓ Adding and Subtracting Polynomials
- ✓ Multiplying Monomials
- ✓ Multiplying and Dividing Monomials
- ✓ Multiplying a Polynomial and a Monomial
- ✓ Multiplying Binomials
- ✓ Factoring Trinomials
- ✓ Operations with Polynomials

Mathematics is the supreme judge; from its decisions there is no appeal. – Tobias Dantzig

HiSET Subject Test – Mathematics

Writing Polynomials in Standard Form

✎ Write each polynomial in standard form.

1) $11x - 7x =$

2) $-5 + 19x - 19x =$

3) $6x^5 - 12x^3 =$

4) $12 + 17x^4 - 12 =$

5) $5x^2 + 4x - 9x^3 =$

6) $-3x^2 + 12x^5 =$

7) $5x + 8x^3 - 2x^8 =$

8) $-7x^3 + 4x - 9x^6 =$

9) $3x^2 + 22 - 6x =$

10) $3 - 4x + 9x^4 =$

11) $13x^2 + 28x - 8x^3 =$

12) $16 + 4x^2 - 2x^3 =$

13) $19x^2 - 9x + 9x^4 =$

14) $3x^4 - 7x^2 - 2x^3 =$

15) $-51 + 3x^2 - 8x^4 =$

16) $7x^2 - 8x^6 + 4x^4 - 15 =$

17) $6x^4 - 4x^5 + 16 - 3x^3 =$

18) $-2x^6 + 4x - 7x^2 - 5x =$

19) $11x^7 + 8x^5 - 5x^7 - 3x^2 =$

20) $2x^2 - 12x^5 + 8x^2 + 3x^6 =$

21) $4x^5 - 11x^7 - 6x^3 + 16x^5 =$

22) $6x^3 + 3x^5 + 34x^4 - 8x^5 =$

23) $3x(4x + 5 - 2x^2) =$

24) $12x(x^6 + 4x^3) =$

25) $5x(3x^2 + 6x + 4) =$

26) $7x(4 - 2x + 6x^5) =$

27) $3x(4x^4 - 4x^3 + 2) =$

28) $4x(2x^5 + 6x^2 - 3) =$

29) $5x(3x^4 + 4x^3 + 2x) =$

30) $2x(3x - 2x^3 + 4x^6) =$

WWW.MathNotion.Com

HiSET Subject Test – Mathematics

Simplifying Polynomials

✏️ **Simplify each expression.**

1) $3(4x - 20) =$

2) $5x(3x - 4) =$

3) $6x(5x - 7) =$

4) $3x(7x + 5) =$

5) $5x(4x - 3) =$

6) $6x(8x + 2) =$

7) $(3x - 2)(x - 4) =$

8) $(x - 5)(2x + 6) =$

9) $(x - 3)(x - 7) =$

10) $(3x + 4)(3x - 4) =$

11) $(5x - 4)(5x - 2) =$

12) $6x^2 + 6x^2 - 8x^4 =$

13) $3x - 2x^2 + 5x^3 + 7 =$

14) $7x + 4x^2 - 10x^3 =$

15) $12x^2 + 5x^5 - 6x^3 =$

16) $-5x^2 + 4x^6 + 6x^8 =$

17) $-12x^3 + 10x^5 - 4x^6 + 4x =$

18) $11 - 7x^2 + 4x^2 - 16x^3 + 11 =$

19) $2x^2 - 9x + 4x^3 + 15x - 10x =$

20) $13 - 7x^5 + 6x^5 - 4x^2 + 5 =$

21) $-5x^8 + x^6 - 14x^3 + 5x^8 =$

22) $(7x^4 - 4) + (7x^4 - 2x^4) =$

23) $3(3x^4 - 4x^3 - 6x^4) =$

24) $-5(x^9 + 8) - 5(10 - x^9) =$

25) $8x^3 - 9x^4 - 2x + 19 - 8x^3 =$

26) $11 - 8x^3 + 6x^3 - 7x^5 + 6 =$

27) $(5x^3 - 4x) - (6x - 2 - 6x^3) =$

28) $4x^2 - 5x^4 - x(3x^3 + 2x) =$

29) $6x + 6x^5 - 10 - 4(x^5 - 3) =$

30) $4 - 3x^4 + (6x^5 - 2x^4 + 5x^5) =$

31) $-(x^5 + 4) - 8(3 + x^5) =$

32) $(4x^3 - 3x) - (3x - 5x^3) =$

WWW.MathNotion.Com

HiSET Subject Test – Mathematics

Adding and Subtracting Polynomials

✎ **Add or subtract expressions.**

1) $(-2x^2 - 3) + (3x^2 + 4) =$

2) $(4x^3 + 6) - (7 - 2x^3) =$

3) $(4x^5 + 5x^2) - (2x^5 + 15) =$

4) $(6x^3 - 2x^2) + (5x^2 - 4x) =$

5) $(10x^4 + 28x) - (34x^4 + 6) =$

6) $(7x^2 - 3) + (7x^2 + 3) =$

7) $(9x^2 + 4) - (10 - 5x^2) =$

8) $(6x^2 + x^5) - (x^5 + 4) =$

9) $(4x^3 - x) + (3x - 7x^3) =$

10) $(11x + 10) - (8x + 10) =$

11) $(15x^3 - 3x) - (3x - 4x^3) =$

12) $(4x - x^5) - (6x^5 + 8x) =$

13) $(2x^2 - 7x^7) - (4x^7 - 6x) =$

14) $(3x^2 - 5) + (8x^2 + 4x^5) =$

15) $(9x^4 + 5x^5) - (x^5 - 9x^4) =$

16) $(-4x^3 - 2x) + (9x - 5x^3) =$

17) $(4x - 3x^2) - (148x^2 + x) =$

18) $(5x - 8x^4) - (3x^4 - 4x^2) =$

19) $(8x^4 - 4) + (2x^4 - 3x^2) =$

20) $(5x^6 + 7x^3) - (x^3 - 5x^6) =$

21) $(-2x^2 + 20x^5 + 5x^4) + (12x^4 + 8x^5 + 24x^2) =$

22) $(7x^4 - 9x^7 - 6x) - (-3x^4 - 9x^7 + 6x) =$

23) $(14x + 12x^4 - 18x^6) + (20x^4 + 18x^6 - 10x) =$

24) $(5x^8 - 6x^6 - 4x) - (5x^3 + 9x^6 - 7x) =$

25) $(11x^2 - 6x^4 - 3x) - (-4x^2 - 12x^4 + 9x) =$

26) $(-5x^9 + 14x^3 + 3x^7) + (10x^7 + 26x^3 + 3x^9) =$

WWW.MathNotion.Com

HiSET Subject Test – Mathematics

Multiplying Monomials

✎ Simplify each expression.

1) $6u^8 \times (-u^2) =$

2) $(-5p^8) \times (-2p^3) =$

3) $4xy^3z^5 \times 3z^4 =$

4) $3u^5t \times 8ut^4 =$

5) $(-5a^2) \times (-7a^3b^6) =$

6) $-3a^4b^3 \times 6a^2b =$

7) $13xy^5 \times x^4y^4 =$

8) $6p^4q^3 \times (-8pq^6) =$

9) $8s^4t^3 \times 4st^3 =$

10) $(-6x^4y^3) \times 6x^2y =$

11) $3xy^7z \times 12z^3 =$

12) $24xy \times x^2y =$

13) $13pq^4 \times (-3p^2q) =$

14) $13s^3t^4 \times st^4 =$

15) $11p^5 \times (-6p^3) =$

16) $(-8p^3q^5r) \times 3pq^4r^6 =$

17) $(-4a^4) \times (-7a^3b) =$

18) $6u^6v^2 \times (-5u^3v^4) =$

19) $9u^5 \times (-3u) =$

20) $-6xy^5 \times 4x^2y =$

21) $13y^5z^3 \times (-y^3z) =$

22) $8a^4bc^3 \times 2abc^3 =$

23) $(-7p^5q^6) \times (-5p^4q^2) =$

24) $4u^5v^3 \times (-4u^7v^3) =$

25) $17y^4z^5 \times (-y^6z) =$

26) $(-5pq^3r^2) \times 8p^2q^4r =$

27) $3ab^5c^6 \times 5a^4bc^2 =$

28) $6x^3yz^2 \times 3x^2y^7z^3 =$

WWW.MathNotion.Com

HiSET Subject Test – Mathematics

Multiplying and Dividing Monomials

✎ Simplify each expression.

1) $(5x^5)(2x^2) =$

2) $(4x^4)(6x^2) =$

3) $(3x^4)(7x^4) =$

4) $(5x^6)(4x^2) =$

5) $(12x^4)(3x^6) =$

6) $(4yx^8)(8y^4x^3) =$

7) $(14x^4y)(x^3y^5) =$

8) $(-5x^3y^4)(2x^3y^5) =$

9) $(-6x^4y^2)(-3x^3y^5) =$

10) $(5x^3y)(-5x^2y^3) =$

11) $(6x^4y^3)(4x^3y^4) =$

12) $(4x^3y^2)(5x^2y^4) =$

13) $(12x^3y^6)(4x^4y^{10}) =$

14) $(15x^3y^5)(3x^4y^6) =$

15) $(7x^2y^7)(8x^6y^7) =$

16) $(-3x^3y^8)(7x^9y^4) =$

17) $\dfrac{5x^6y^6}{xy^4} =$

18) $\dfrac{19x^7y^5}{19x^6y} =$

19) $\dfrac{56x^4y^4}{8xy} =$

20) $\dfrac{81x^5y^6}{9x^4y^5} =$

21) $\dfrac{36x^7y^6}{9x^2y^3} =$

22) $\dfrac{48x^9y^7}{4x^4y^6} =$

23) $\dfrac{88x^{18}y^{12}}{11x^8y^9} =$

24) $\dfrac{30x^7y^6}{6x^8y^3} =$

25) $\dfrac{150x^7y^6}{30x^4y^6} =$

26) $\dfrac{-42x^{18}y^{14}}{6x^4y^9} =$

27) $\dfrac{-36x^7y^8}{9x^5y^8} =$

WWW.MathNotion.Com

HiSET Subject Test – Mathematics

Multiplying a Polynomial and a Monomial

✏ **Find each product.**

1) $x(2x + 4) =$

2) $6(4 - 2x) =$

3) $5x(4x + 2) =$

4) $x(-4x + 5) =$

5) $8x(2x - 2) =$

6) $6(2x - 4y) =$

7) $7x(5x - 5) =$

8) $3x(12x + 2y) =$

9) $4x(x + 6y) =$

10) $11x(3x + 4y) =$

11) $7x(3x + 2) =$

12) $10x(4x - 10y) =$

13) $9x(3x - 2y) =$

14) $7x(x - 4y + 6) =$

15) $8x(2x^2 + 5y^2) =$

16) $12x(2x + 3y) =$

17) $4(2x^4 - 4y^4) =$

18) $4x(-3x^2y + 4y) =$

19) $-4(5x^3 - 2xy + 4) =$

20) $4(x^2 - 5xy - 6) =$

21) $8x(2x^3 - 5xy + 2x) =$

22) $-6x(-2x^3 - 6x + 2xy) =$

23) $3(2x^2 + xy - 9y^2) =$

24) $4x(5x^3 - 3x + 7) =$

25) $6(3x^{22} - 2x - 5) =$

26) $x^2(-2x^3 + 4x + 3) =$

27) $x^2(4x^3 + 10 - 2x) =$

28) $4x^4(3x^3 - 2x + 5) =$

29) $2x^2(4x^4 - 5xy + 7y^3) =$

30) $5x^2(5x^4 - 3x + 9) =$

31) $7x^2(6x^2 + 3x - 6) =$

32) $4x(x^3 - 4xy + 2y^2) =$

WWW.MathNotion.Com

HiSET Subject Test – Mathematics

Multiplying Binomials

✎ **Find each product.**

1) $(x + 3)(x + 6) =$

2) $(x - 4)(x + 3) =$

3) $(x - 3)(x - 8) =$

4) $(x + 8)(x + 9) =$

5) $(x - 2)(x - 12) =$

6) $(x + 5)(x + 5) =$

7) $(x - 6)(x + 7) =$

8) $(x - 8)(x - 3) =$

9) $(x + 7)(x + 12) =$

10) $(x - 4)(x + 8) =$

11) $(x + 8)(x + 8) =$

12) $(x + 2)(x + 7) =$

13) $(x - 6)(x + 6) =$

14) $(x - 5)(x + 5) =$

15) $(x + 11)(x + 11) =$

16) $(x + 6)(x + 9) =$

17) $(x - 2)(x + 2) =$

18) $(x - 4)(x + 7) =$

19) $(3x + 5)(x + 6) =$

20) $(5x - 6)(4x + 8) =$

21) $(x - 7)(3x + 7) =$

22) $(x - 9)(x - 4) =$

23) $(x - 12)(x + 2) =$

24) $(2x - 4)(5x + 4) =$

25) $(3x - 8)(x + 8) =$

26) $(7x - 2)(6x + 3) =$

27) $(4x + 5)(3x + 5) =$

28) $(7x - 4)(9x + 4) =$

29) $(x + 2)(2x - 8) =$

30) $(5x - 4)(5x + 4) =$

31) $(3x + 2)(3x - 7) =$

32) $(x^2 + 8)(x^2 - 8) =$

WWW.MathNotion.Com

HiSET Subject Test – Mathematics

Factoring Trinomials

✍ **Factor each trinomial.**

1) $x^2 + 8x + 12 =$

2) $x^2 - 6x + 5 =$

3) $x^2 + 15x + 36 =$

4) $x^2 - 12x + 35 =$

5) $x^2 - 11x + 18 =$

6) $x^2 - 9x + 18 =$

7) $x^2 + 18x + 72 =$

8) $x^2 - x - 72 =$

9) $x^2 + 4x - 21 =$

10) $x^2 - 13x + 22 =$

11) $x^2 + 2x - 24 =$

12) $x^2 - 3x - 40 =$

13) $x^2 - 3x - 70 =$

14) $x^2 + 26x + 169 =$

15) $4x^2 - 7x - 15 =$

16) $x^2 - 14x + 33 =$

17) $10x^2 + 5x - 15 =$

18) $6x^2 - 4x - 42 =$

19) $x^2 + 12x + 36 =$

20) $5x^2 + 17x - 12 =$

✍ **Calculate each problem.**

21) The area of a rectangle is $x^2 - x - 56$. If the width of rectangle is $x + 7$, what is its length? _____

22) The area of a parallelogram is $4x^2 + 17x - 15$ and its height is $x + 5$. What is the base of the parallelogram? _____

23) The area of a rectangle is $6x^2 - 22x + 12$. If the width of the rectangle is $3x - 2$, what is its length? _____

WWW.MathNotion.Com

HiSET Subject Test – Mathematics

Operations with Polynomials

✎ **Find each product.**

1) $4(5x + 3) =$ _____

2) $8(2x + 6) =$ _____

3) $2(5x - 2) =$ _____

4) $-4(7x - 3) =$ _____

5) $3x^2(9x + 1) =$ _____

6) $4x^6(7x - 9) =$ _____

7) $3x^4(-7x + 3) =$ _____

8) $-8x^4(5x - 8) =$ _____

9) $7(x^2 + 5x - 3) =$ _____

10) $9(5x^2 - 7x + 5) =$ _____

11) $3(3x^2 + 3x + 2) =$ _____

12) $5x(3x^2 + 5x + 8) =$ _____

13) $(5x + 7)(3x - 3) =$ _____

14) $(9x + 3)(3x - 5) =$ _____

15) $(6x + 3)(4x - 2) =$ _____

16) $(7x - 2)(3x + 5) =$ _____

✎ **Calculate each problem.**

17) The measures of two sides of a triangle are $(2x + 5y)$ and $(6x - 3y)$. If the perimeter of the triangle is $(13x + 4y)$, what is the measure of the third side? _____

18) The height of a triangle is $(8x + 5)$ and its base is $(4x - 3)$. What is the area of the triangle? _____

19) One side of a square is $(6x + 2)$. What is the area of the square? _____

20) The length of a rectangle is $(5x - 8y)$ and its width is $(15x + 8y)$. What is the perimeter of the rectangle? _____

21) The side of a cube measures $(x + 2)$. What is the volume of the cube? _____

22) If the perimeter of a rectangle is $(28x + 6y)$ and its width is $(5x + 2y)$, what is the length of the rectangle? _____

WWW.MathNotion.Com

HiSET Subject Test – Mathematics

Answers of Worksheets

Writing Polynomials in Standard Form

1) $4x$
2) -5
3) $6x^5 - 12x^3$
4) $14x^4$
5) $-9x^3 + 5x^2 + 4x$
6) $12x^5 - 3x^2$
7) $-2x^8 + 8x^3 + 5x$
8) $-9x^6 - 7x^3 + 4x$
9) $3x^2 - 6x + 22$
10) $9x^4 - 4x + 3$
11) $-8x^3 + 13x^2 + 28x$
12) $-2x^3 + 4x^2 + 16$
13) $9x^4 + 19x^2 - 9x$
14) $3x^4 - 2x^3 - 7x^2$
15) $-8x^4 + 3x^2 - 51$

16) $-8x^6 + 4x^4 + 7x^2 - 15$
17) $-4x^5 + 6x^4 - 3x^3 + 16$
18) $-2x^6 - 7x^2 - x$
19) $6x^7 + 8x^5 - 3x^2$
20) $3x^6 - 12x^5 + 10x^2$
21) $-11x^7 + 20x^5 - 6x^3$
22) $-5x^5 + 34x^4 + 6x^3$
23) $-6x^3 + 12x^2 + 15x$
24) $12x^7 + 48x^4$
25) $15x^3 + 30x^2 + 20x$
26) $42x^6 - 14x^2 + 28x$
27) $12x^5 - 12x^4 + 6x$
28) $8x^6 + 24x^3 - 12x$
29) $15x^5 + 20x^4 + 10x^2$
30) $8x^7 - 4x^4 + 6x^2$

Simplifying Polynomials

1) $12x - 60$
2) $15x^2 - 20x$
3) $30x^2 - 42x$
4) $21x^2 + 15x$
5) $20x^2 - 15x$
6) $48x^2 + 12x$
7) $3x^2 - 14x + 8$
8) $2x^2 - 4x - 30$
9) $x^2 - 10x + 21$
10) $9x^2 - 16$
11) $25x^2 - 30x + 8$
12) $-8x^4 + 12x^2$

13) $5x^3 - 2x^2 + 3x + 7$
14) $-10x^3 + 4x^2 + 7x$
15) $5x^5 - 6x^3 + 12x^2$
16) $6x^8 + 4x^6 - 5x^2$
17) $-4x^6 + 10x^5 - 12x^3 + 4x$
18) $-16x^3 - 3x^2 + 22$
19) $4x^3 + 2x^2 - 4x$
20) $-x^5 - 4x^2 + 18$
21) $x^6 - 14x^3$
22) $12x^4 - 4$
23) $-9x^4 - 12x^3$
24) -90

WWW.MathNotion.Com

HiSET Subject Test – Mathematics

25) $-9x^4 - 2x + 19$

26) $-7x^5 - 2x^3 + 17$

27) $11x^3 - 10x + 2$

28) $-8x^4 + 2x^2$

29) $2x^5 + 6x + 2$

30) $11x^5 - 5x^4 + 4$

31) $-9x^5 - 28$

32) $9x^3 - 6x$

Adding and Subtracting Polynomials

1) $x^2 + 1$

2) $6x^3 - 1$

3) $2x^5 + 5x^2 - 15$

4) $6x^3 + 3x^2 - 4x$

5) $-24x^4 + 28x - 6$

6) $14x^2$

7) $14x^2 - 6$

8) $6x^2 - 4$

9) $-3x^3 + 2x$

10) $3x$

11) $19x^3 - 6x$

12) $-7x^5 - 4x$

13) $-11x^7 + 2x^2 + 6x$

14) $4x^5 + 11x^2 - 5$

15) $4x^5 + 18x^4$

16) $-9x^3 + 7x$

17) $-151x^2 + 3x$

18) $-11x^4 + 4x^2 + 5x$

19) $10x^4 - 3x^2 - 4$

20) $10x^6 + 6x^3$

21) $28x^5 + 17x^4 + 22x^2$

22) $10x^4 - 12x$

23) $32x^4 + 4x$

24) $5x^8 - 15x^6 - 5x^3 + 3x$

25) $6x^4 + 15x^2 - 12x$

26) $-2x^9 + 13x^7 + 40x^3$

Multiplying Monomials

1) $-6u^{10}$

2) $10p^{11}$

3) $12xy^3z^9$

4) $24u^6t^5$

5) $35a^5b^6$

6) $-18a^6b^4$

7) $13x^5y^9$

8) $-48p^5q^9$

9) $32s^5t^6$

10) $-36x^6y^4$

11) $36xy^7z^4$

12) $24px^3y^2$

13) $-39p^3q^5$

14) $13s^4t^8$

15) $-66p^8$

16) $-24p^4q^9r^7$

17) $28a^7b$

18) $-30u^9v^6$

19) $-27u^6$

20) $-24x^3y^6$

21) $-13y^8z^4$

22) $16a^5b^2c^6$

23) $35p^9q^8$

24) $-16u^{12}v^6$

25) $-17y^{10}z^6$

26) $-40p^3q^7r^3$

27) $15a^5b^6c^8$

28) $18x^5y^8z^5$

Multiplying and Dividing Monomials

1) $10x^7$

2) $24x^6$

3) $21x^8$

4) $20x^8$

5) $36x^{10}$

6) $32x^{11}y^5$

7) $14x^7y^6$

8) $-10x^6y^9$

9) $18x^7y^7$

10) $-25x^5y^4$

11) $24x^7y^7$

12) $20x^5y^6$

HiSET Subject Test – Mathematics

13) $48x^7y^{16}$
14) $45x^7y^{11}$
15) $56x^8y^{14}$
16) $-21x^{12}y^{12}$
17) $5x^5y^2$

18) xy^4
19) $7x^3y^3$
20) $9xy$
21) $4x^5y^3$
22) $12x^5y$

23) $8x^{10}y^3$
24) $5x^{-1}y^3$
25) $5x^3$
26) $-7x^{14}y^5$
27) $-4x^2$

Multiplying a Polynomial and a Monomial

1) $2x^2 + 4x$
2) $-12x + 24$
3) $20x^2 + 10x$
4) $-4x^2 + 5x$
5) $16x^2 - 16x$
6) $12x - 24y$
7) $35x^2 - 35x$
8) $36x^2 + 6xy$
9) $4x^2 + 24xy$
10) $33x^2 + 44xy$
11) $21x^2 + 14x$
12) $40x^2 - 100xy$
13) $27x^2 - 18xy$
14) $7x^2 - 28xy + 42x$
15) $16x^3 + 40xy^2$
16) $24x^2 + 36xy$

17) $8x^4 - 16y^4$
18) $-12x^3y + 16xy$
19) $-20x^3 + 8xy - 16$
20) $4x^2 - 20xy - 24$
21) $16x^4 - 40x^2y + 16x^2$
22) $12x^4 + 36x^2 - 12x^2y$
23) $6x^2 + 3xy - 27y^2$
24) $20x^4 - 12x^2 + 28x$
25) $18x^{22} - 12x - 30$
26) $-2x^5 + 4x^3 + 3x^2$
27) $4x^5 - 2x^3 + 10x^2$
28) $12x^7 - 8x^5 + 20x^4$
29) $8x^6 - 10x^3y + 14x^2y^3$
30) $25x^6 - 15x^3 + 45x^2$
31) $42x^4 + 21x^3 - 42x^2$
32) $4x^4 - 16x^2y + 8xy^2$

Multiplying Binomials

1) $x^2 + 9x + 18$
2) $x^2 - x - 12$
3) $x^2 - 11x + 24$
4) $x^2 + 17x + 72$
5) $x^2 - 14x + 24$
6) $x^2 + 10x + 25$
7) $x^2 + x - 42$

8) $x^2 - 11x + 24$
9) $x^2 + 19x + 84$
10) $x^2 + 4x - 32$
11) $x^2 + 16x + 64$
12) $x^2 + 9x + 14$
13) $x^2 - 36$
14) $x^2 - 25$

HiSET Subject Test – Mathematics

15) $x^2 + 22x + 121$
16) $x^2 + 15x + 54$
17) $x^2 - 4$
18) $x^2 + 3x - 28$
19) $3x^2 + 23x + 30$
20) $20x^2 + 16x - 48$
21) $3x^2 - 14x - 49$
22) $x^2 - 13x + 36$
23) $x^2 - 10x - 24$

24) $10x^2 - 12x - 16$
25) $3x^2 + 16x - 64$
26) $42x^2 + 9x - 6$
27) $12x^2 + 35x + 25$
28) $63x^2 - 8x - 16$
29) $2x^2 - 4x - 16$
30) $25x^2 - 16$
31) $9x^2 - 15x - 14$
32) $x^4 - 64$

Factoring Trinomials

1) $(x + 6)(x + 2)$
2) $(x - 5)(x - 1)$
3) $(x + 12)(x + 3)$
4) $(x - 5)(x - 7)$
5) $(x - 2)(x - 9)$
6) $(x - 6)(x - 3)$
7) $(x + 6)(x + 12)$
8) $(x + 8)(x - 9)$

9) $(x - 3)(x + 7)$
10) $(x - 11)(x - 2)$
11) $(x - 4)(x + 6)$
12) $(x - 8)(x + 5)$
13) $(x + 7)(x - 10)$
14) $(x + 13)(x + 13)$
15) $(4x + 5)(x - 3)$
16) $(x - 11)(x - 3)$

17) $(5x - 5)(2x + 3)$
18) $(2x - 6)(3x + 7)$
19) $(x + 6)(x + 6)$
20) $(5x - 3)(x + 4)$
21) $(x - 8)$
22) $(4x - 3)$
23) $(2x - 6)$

Operations with Polynomials

1) $20x + 12$
2) $16x + 48$
3) $10x - 4$
4) $-28x + 12$
5) $27x^3 + 3x^2$
6) $28x^7 - 36x^6$
7) $-21x^5 + 9x^4$
8) $-40x^5 + 64x^4$

9) $7x^2 + 35x - 21$
10) $45x^2 - 63x + 45$
11) $9x^2 + 9x + 6$
12) $15x^3 + 25x^2 + 40x$
13) $15x^2 + 6x - 21$
14) $27x^2 - 36x - 15$
15) $24x^2 - 6$
16) $21x^2 + 29x - 10$

17) $(5x + 2y)$
18) $16x^2 - 2x - \frac{15}{2}$
19) $36x^2 + 24x + 4$
20) $40x$
21) $x^3 + 6x^2 + 12x + 8$
22) $(9x + y)$

WWW.MathNotion.Com

HiSET Subject Test – Mathematics

Chapter 9 : Functions Operations and Quadratic

Topics that you'll practice in this chapter:

- ✓ Evaluating Function
- ✓ Adding and Subtracting Functions
- ✓ Multiplying and Dividing Functions
- ✓ Composition of Functions
- ✓ Quadratic Equation
- ✓ Solving Quadratic Equations
- ✓ Quadratic Formula and the Discriminant
- ✓ Graphing Quadratic Functions

It's fine to work on any problem, so long as it generates interesting mathematics along the way – even if you don't solve it at the end of the day." – Andrew Wiles

HiSET Subject Test – Mathematics

Evaluating Function

✎ **Write each of following in function notation.**

1) $h = -8x + 3$

2) $k = 2a - 14$

3) $d = 11t$

4) $y = \frac{5}{12}x - \frac{7}{12}$

5) $m = 24n - 210$

6) $c = p^2 - 5p + 10$

✎ **Evaluate each function.**

7) $f(x) = 2x - 7$, find $f(-3)$

8) $g(x) = \frac{1}{9}x + 12$, find $f(18)$

9) $h(x) = -4x + 9$, find $f(3)$

10) $f(x) = -x + 19$, find $f(-3)$

11) $f(a) = 7a - 12$, find $f(3)$

12) $h(x) = 14 - 3x$, find $f(-4)$

13) $g(n) = 6n - 10$, find $f(2)$

14) $f(x) = -11x - 4$, find $f(-1)$

15) $k(n) = -20 - 3.5n$, find $f(2)$

16) $f(x) = -0.7x + 3.3$, find $f(-7)$

17) $g(n) = \frac{11n+8}{n}$, find $g(2)$

18) $g(n) = \sqrt{3n} + 12$, find $g(3)$

19) $h(x) = x^{-2} - 7$, find $h(\frac{1}{9})$

20) $h(n) = n^{-3} + 11$, find $h(\frac{1}{4})$

21) $h(n) = n^3 - 2$, find $h(\frac{1}{2})$

22) $h(n) = n^2 - 4$, find $h(-\frac{1}{3})$

23) $h(n) = 4n^2 - 13$, find $h(-5)$

24) $h(n) = -2n^3 - 6n$, find $h(2)$

25) $g(n) = \sqrt{16n^2} - \sqrt{n}$, find $g(4)$

26) $h(a) = \frac{-14a+9}{3a}$, find $h(-b)$

27) $k(a) = 12a - 14$, find $k(a - 3)$

28) $h(x) = \frac{1}{9}x + 18$, find $h(-18x)$

29) $h(x) = 8x^2 + 16$, find $h(\frac{x}{2})$

30) $h(x) = x^4 - 20$, find $h(-2x)$

HiSET Subject Test – Mathematics

Adding and Subtracting Functions

✏️ **Perform the indicated operation.**

1) $f(x) = 2x + 3$

   $g(x) = x + 7$

   Find $(f - g)(2)$

2) $g(a) = -5a - 8$

   $f(a) = -3a - 5$

   Find $(g - f)(-2)$

3) $h(t) = 4t + 3$

   $g(t) = 4t + 7$

   Find $(h - g)(t)$

4) $g(a) = -6a - 10$

   $f(a) = 3a^2 + 9$

   Find $(g - f)(x)$

5) $g(x) = \frac{5}{6}x - 23$

   $h(x) = \frac{5}{12}x + 25$

   Find $g(12) - h(12)$

6) $h(x) = \sqrt{3x} - 2$

   $g(x) = \sqrt{3x} + 5$

   Find $(h + g)(12)$

7) $f(x) = x^{-1}$

   $g(x) = x^2 + \frac{5}{x}$

   Find $(f - g)(-3)$

8) $h(n) = n^2 + 2$

   $g(n) = -4n + 6$

   Find $(h - g)(2a)$

9) $g(x) = -2x^2 - 5 - 4x$

   $f(x) = 7 + 2x$

   Find $(g - f)(3x)$

10) $g(t) = 11t - 4$

    $f(t) = -2t^2 + 5$

    Find $(g + f)(-t)$

11) $f(x) = 8x + 9$

    $g(x) = -5x^2 + 3x$

    Find $(f - g)(-x^2)$

12) $f(x) = -3x^4 - 5x$

    $g(x) = 2x^4 + 5x + 22$

    Find $(f + g)(3x^2)$

WWW.MathNotion.Com

HiSET Subject Test – Mathematics

Multiplying and Dividing Functions

✎ **Perform the indicated operation.**

1) $g(x) = -2x - 1$
   $f(x) = 4x + 3$
   Find $(g.f)(2)$

2) $f(x) = 5x$
   $h(x) = -2x + 3$
   Find $(f.h)(-2)$

3) $g(a) = 5a - 2$
   $h(a) = 2a - 3$
   Find $(g.h)(-3)$

4) $f(x) = 2x - 7$
   $h(x) = x - 5$
   Find $\left(\frac{f}{h}\right)(4)$

5) $f(x) = 8a^2$
   $g(x) = 3 + 2a$
   Find $\left(\frac{f}{g}\right)(2)$

6) $g(a) = \sqrt{4a} + 2$
   $f(a) = (-a)^4 + 1$
   Find $\left(\frac{g}{f}\right)(1)$

7) $g(t) = t^3 + 1$
   $h(t) = 5t - 2$
   Find $(g.h)(-2)$

8) $g(n) = n^2 + 2n - 4$
   $h(n) = -5n + 3$
   Find $(g.h)(1)$

9) $g(a) = (a - 3)^2$
   $f(a) = a^2 + 4$
   Find $\left(\frac{g}{f}\right)(3)$

10) $g(x) = -3x^2 + \frac{4}{5}x + 9$
    $f(x) = x^2 - 24$
    Find $\left(\frac{g}{f}\right)(5)$

11) $f(x) = 2x^3 - 5x^2 + 1$
    $g(x) = 3x - 1$
    Find $(f.g)(x)$

12) $f(x) = 5x - 2$
    $g(x) = x^3 - 2x$
    Find $(f.g)(x^2)$

WWW.MathNotion.Com

Composition of Functions

Using $f(x) = 2x - 5$ and $g(x) = -2x$, find:

1) $f(g(2)) =$

2) $f(g(-1)) =$

3) $g(f(-4)) =$

4) $g(f(5)) =$

5) $f(g(3)) =$

6) $g(f(0)) =$

Using $f(x) = -\frac{1}{4}x + \frac{3}{4}$ and $g(x) = 2x^2$, find:

7) $g(f(-2)) =$

8) $g(f(4)) =$

9) $g(g(1)) =$

10) $f(f(1)) =$

11) $g(f(-4)) =$

12) $f(g(x)) =$

Using $f(x) = -2x + 2$ and $g(x) = x + 1$, find:

13) $g(f(1)) =$

14) $f(f(0)) =$

15) $f(g(-1)) =$

16) $f(g(-3)) =$

17) $g(f(2)) =$

18) $f(g(x)) =$

Using $f(x) = \sqrt{x+9}$ and $g(x) = x - 9$, find:

19) $f(g(9)) =$

20) $g(f(-9)) =$

21) $f(g(4)) =$

22) $f(f(7)) =$

23) $g(f(-5)) =$

24) $g(g(0)) =$

HiSET Subject Test – Mathematics

Quadratic Equation

✎ **Multiply.**

1) $(x - 4)(x + 6) =$ _____

2) $(x + 5)(x + 7) =$ _____

3) $(x - 6)(x + 8) =$ _____

4) $(x + 2)(x - 9) =$ _____

5) $(x - 7)(x - 8) =$ _____

6) $(3x + 2)(x - 3) =$ _____

7) $(4x - 3)(x + 2) =$ _____

8) $(4x - 5)(x + 1) =$ _____

9) $(7x + 1)(x - 6) =$ _____

10) $(5x + 1)(3x - 3) =$ _____

✎ **Factor each expression.**

11) $x^2 - 2x - 8 =$ _____

12) $x^2 + 8x + 15 =$ _____

13) $x^2 - 2x - 24 =$ _____

14) $x^2 - 10x + 21 =$ _____

15) $x^2 + 10x + 21 =$ _____

16) $4x^2 + 9x + 5 =$ _____

17) $5x^2 + 13x - 6 =$ _____

18) $5x^2 + 17x - 12 =$ _____

19) $2x^2 + 7x + 5 =$ _____

20) $9x^2 - 21x + 6 =$ _____

✎ **Calculate each equation.**

21) $(x + 6)(x - 3) = 0$

22) $(x + 1)(x + 8) = 0$

23) $(3x + 6)(x + 5) = 0$

24) $(2x - 2)(4x + 8) = 0$

25) $x^2 + x + 10 = 22$

26) $x^2 + 11x + 36 = 12$

27) $2x^2 + 9x + 9 = 5$

28) $x^2 + 3x - 24 = 4$

29) $5x^2 + 5x - 40 = 20$

30) $8x^2 + 8x = 48$

WWW.MathNotion.Com

HiSET Subject Test – Mathematics

Solving Quadratic Equations

✎ **Solve each equation by factoring or using the quadratic formula.**

1) $(x + 9)(x - 1) = 0$

2) $(x + 7)(x + 6) = 0$

3) $(x - 8)(x + 3) = 0$

4) $(x - 6)(x - 4) = 0$

5) $(x + 2)(x + 12) = 0$

6) $(5x + 4)(x + 7) = 0$

7) $(6x + 1)(4x + 5) = 0$

8) $(2x + 7)(x + 8) = 0$

9) $(x + 6)(3x + 15) = 0$

10) $(12x + 2)(x + 8) = 0$

11) $x^2 = 8x$

12) $x^2 - 16 = 0$

13) $3x^2 + 6 = 9x$

14) $-2x^2 - 8 = 10x$

15) $5x^2 + 40x = 45$

16) $x^2 + 10x = 24$

17) $x^2 + 6x = 16$

18) $x^2 + 9x = -18$

19) $x^2 + 13x = -36$

20) $x^2 + 3x - 15 = 5x$

21) $x^2 + 8x + 7 = -8$

22) $3x^2 - 11x = -9 + x$

23) $10x^2 + 3 = 27x - 15$

24) $7x^2 - 6x + 8 = 8$

25) $2x^2 - 12 = -3x + 2$

26) $10x^2 - 26x - 3 = -15$

27) $3x^2 + 21 = -16x + 5$

28) $x^2 + 15x - 10 = -66$

29) $3x^2 - 8x - 8 = 4 + x$

30) $2x^2 + 6x - 24 = 12$

31) $3x^2 - 33x + 54 = -18$

32) $-10x^2 - 15x - 9 = -9 - 27x^2$

WWW.MathNotion.Com

HiSET Subject Test – Mathematics

Quadratic Formula and the Discriminant

✏ **Find the value of the discriminant of each quadratic equation.**

1) $3x(x - 8) = 0$

2) $2x^2 + 6x - 4 = 0$

3) $x^2 + 6x + 7 = 0$

4) $x^2 - x + 3 = 0$

5) $x^2 + 4x - 3 = 0$

6) $2x^2 + 6x - 10 = 0$

7) $3x^2 + 7x + 5 = 0$

8) $x^2 - 6x - 4 = 0$

9) $2x^2 + 8x + 3 = 0$

10) $x^2 + 7x - 5 = 0$

11) $5x^2 + 2x - 3 = 0$

12) $-3x^2 - 11x + 4 = 0$

13) $-6x^2 - 12x + 8 = 0$

14) $-x^2 - 9x - 12 = 0$

15) $7x^2 - 6x - 10 = 0$

16) $-4x^2 - 2x + 8 = 0$

17) $5x^2 + 8x - 2 = 0$

18) $6x^2 - 4x = 0$

19) $3x^2 - 5x + 2 = 0$

20) $4x^2 + 9x + 3 = 0$

✏ **Find the discriminant of each quadratic equation then state the number of real and imaginary solutions.**

21) $-4x^2 - 16 = 16x$

22) $20x^2 = 20x - 5$

23) $-11x^2 - 19x = 26$

24) $22x^2 - 4x + 1 = 18x^2$

25) $-11x^2 = -15x + 8$

26) $3x^2 + 6x + 9 = 6$

27) $13x^2 - 5x - 12 = -26$

28) $-8x^2 - 32x - 25 = 7$

WWW.MathNotion.Com

Graphing Quadratic Functions

✎ Sketch the graph of each function. Identify the vertex and axis of symmetry.

1) $y = (x + 3)^2 + 2$

2) $y = (x - 3)^2 - 2$

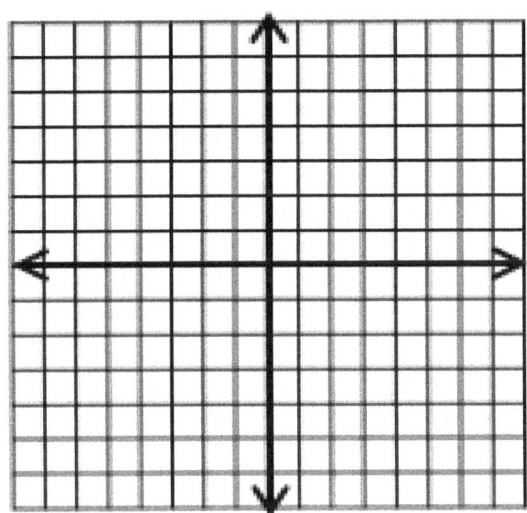 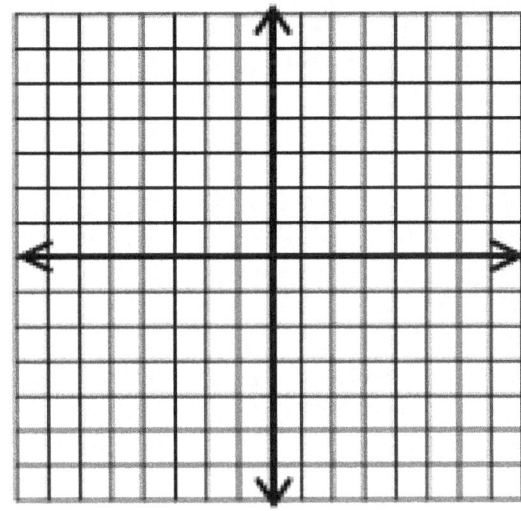

3) $y = 6 - (-x + 4)^2$

4) $y = -3x^2 - 6x + 9$

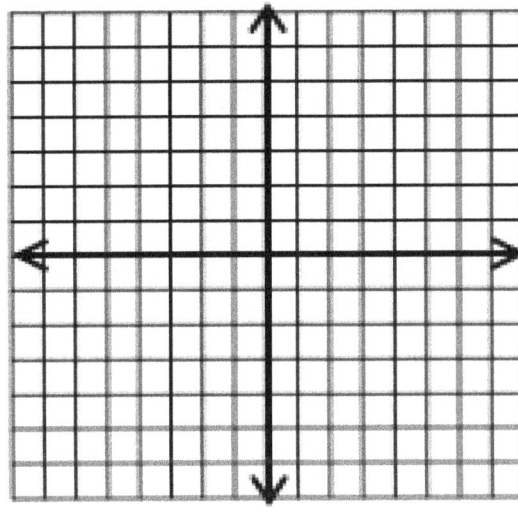 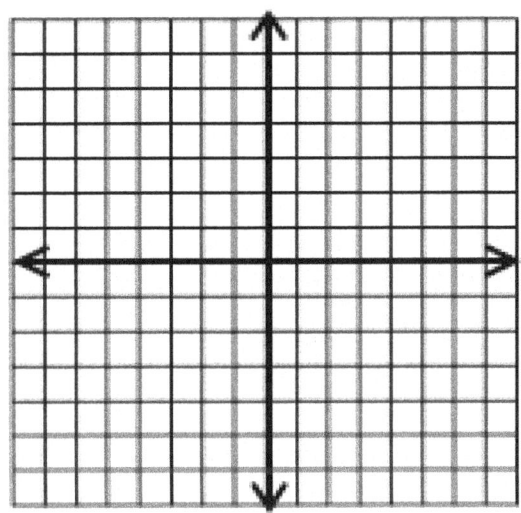

HiSET Subject Test – Mathematics

Answers of Worksheets

Evaluating Function

1) $h(x) = -8x + 3$
2) $k(a) = 2a - 14$
3) $d(t) = 11t$
4) $f(x) = \frac{5}{12}x - \frac{7}{12}$
5) $m(n) = 24n - 210$
6) $c(p) = p^2 - 5p + 10$
7) -13
8) 14
9) -3
10) 22
11) 9
12) 26
13) 2
14) 7
15) -27
16) 8.2
17) 15
18) 15
19) 74
20) 75
21) $-1\frac{7}{8}$
22) $-3\frac{8}{9}$
23) 87
24) -28
25) 14
26) $-\frac{14b+9}{3b}$
27) $12a - 50$
28) $-2x + 18$
29) $2x^2 + 16$
30) $16x^4 - 20$

Adding and Subtracting Functions

1) -2
2) 1
3) -4
4) $-3x^2 - 6x - 19$
5) -43
6) 15
7) $-7\frac{2}{3}$
8) $4a^2 + 8a - 4$
9) $-18x^2 - 18x - 12$
10) $-2t^2 - 11t + 1$
11) $5x^4 - 5x^2 + 9$
12) $-81x^8 + 22$

Multiplying and Dividing Functions

1) -55
2) -70
3) 153
4) -1
5) $4\frac{4}{7}$
6) 2
7) 84
8) 2
9) 0
10) -62
11) $6x^4 - 17x^3 + 5x^2 + 3x - 1$
12) $5x^8 - 2x^6 - 10x^4 + 4x^2$

Composition of Functions

1) -13
2) -1
3) 26
4) -10
5) -17
6) 10
7) $\frac{25}{8}$
8) $\frac{1}{8}$
9) 8
10) $\frac{5}{8}$
11) $\frac{49}{8}$
12) $-\frac{1}{2}(x^2 - \frac{3}{2})$
13) 1
14) -2
15) 2
16) 6
17) -1
18) $-2x$

WWW.MathNotion.Com

HiSET Subject Test – Mathematics

19) 3 21) 2 23) −7
20) −9 22) $\sqrt{13}$ 24) −18

Quadratic Equations

1) $x^2 + 2x - 24$
2) $x^2 + 12x + 35$
3) $x^2 + 2x - 48$
4) $x^2 - 7x - 18$
5) $x^2 - 15x + 56$
6) $3x^2 - 7x - 6$
7) $4x^2 + 5x - 6$
8) $4x^2 - x - 5$
9) $7x^2 - 41x - 6$
10) $15x^2 - 12x - 3$

11) $(x - 4)(x + 2)$
12) $(x + 5)(x + 3)$
13) $(x - 6)(x + 4)$
14) $(x - 3)(x - 7)$
15) $(x + 3)(x + 7)$
16) $(4x + 5)(x + 1)$
17) $(5x - 2)(x + 3)$
18) $(5x - 3)(x + 4)$
19) $(2x + 5)(x + 1)$
20) $3(x - 2)(3x - 1)$

21) $x = -6, x = 3$
22) $x = -1, x = -8$
23) $x = -2, x = -5$
24) $x = 1, x = -2$
25) $x = 3, x = -4$
26) $x = -3, x = -8$
27) $x = -4, x = -\frac{1}{2}$
28) $x = 4, x = -7$
29) $x = 3, x = -4$
30) $x = -3, x = 2$

Solving quadratic equations

1) $\{-9, 1\}$
2) $\{-6, -7\}$
3) $\{8, -3\}$
4) $\{6, 4\}$
5) $\{-2, -12\}$
6) $\{-\frac{4}{5}, -7\}$
7) $\{-\frac{5}{4}, -\frac{1}{6}\}$
8) $\{-\frac{7}{2}, -8\}$

9) $\{-6, -5\}$
10) $\{-\frac{1}{6}, -8\}$
11) $\{8, 0\}$
12) $\{4, -4\}$
13) $\{2, 1\}$
14) $\{-4, -1\}$
15) $\{1, -9\}$
16) $\{2, -12\}$

17) $\{2, -8\}$
18) $\{-3, -6\}$
19) $\{-4, -9\}$
20) $\{5, -3\}$
21) $\{-5, -3\}$
22) $\{1, 3\}$
23) $\{\frac{6}{5}, \frac{3}{2}\}$
24) $\{\frac{6}{7}, 0\}$

25) $\{-\frac{7}{2}, 2\}$
26) $\{\frac{3}{5}, 2\}$
27) $\{-\frac{4}{3}, -4\}$
28) $\{-8, -7\}$
29) $\{4, -1\}$
30) $\{3, -6\}$
31) $\{3, 8\}$
32) $\{\frac{15}{17}, 0\}$

Quadratic formula and the discriminant

1) 576
2) 68
3) 8
4) −11
5) 28

6) 116
7) −11
8) 52
9) 40
10) 69

11) 64
12) 169
13) 336
14) 33
15) 316

16) 132
17) 104
18) 16
19) 1
20) 33

21) 0, *one real solution* 22) 0, *one real solution* 23) −783, *no solution*

WWW.MathNotion.Com

HiSET Subject Test – Mathematics

24) 0, *one real solution* 26) 0, *one real solution* 28) 0, *one real solution*

25) −127, *no solution* 27) −703, *no solution*

Graphing quadratic functions

1) $(-3, 2), x = -3$

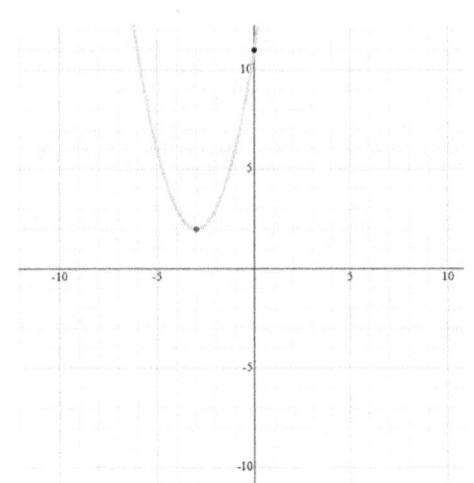

2) $(3, -2), x = 3$

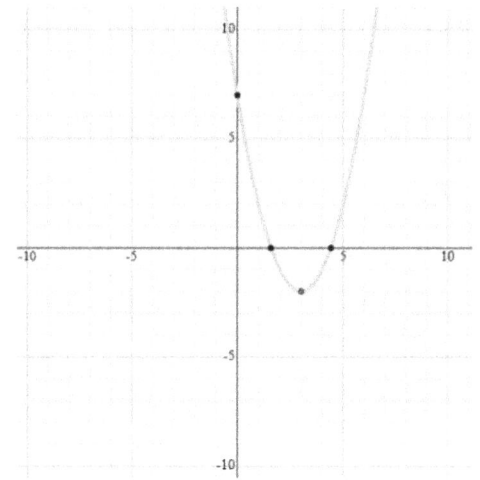

3) $(4, 6), x = 4$

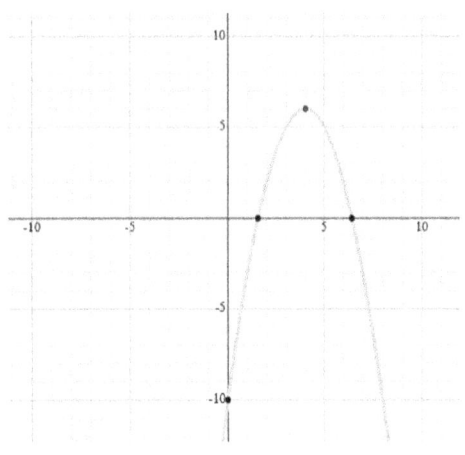

4) $(-1, 12), x = -1$

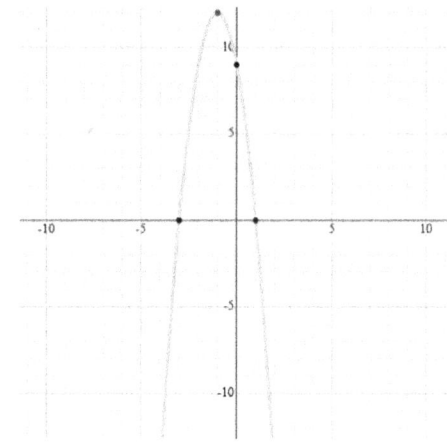

HiSET Subject Test – Mathematics

Chapter 10 :
Geometry and Solid Figures

Topics that you'll practice in this chapter:

- ✓ Angles
- ✓ Pythagorean Relationship
- ✓ Triangles
- ✓ Polygons
- ✓ Trapezoids
- ✓ Circles
- ✓ Cubes
- ✓ Rectangular Prism
- ✓ Cylinder
- ✓ Pyramids and Cone

Mathematics is, as it were, a sensuous logic, and relates to philosophy as do the arts, music, and plastic art to poetry. — K. Shegel

HiSET Subject Test – Mathematics

Angles

✎ **What is the value of *x* in the following figures?**

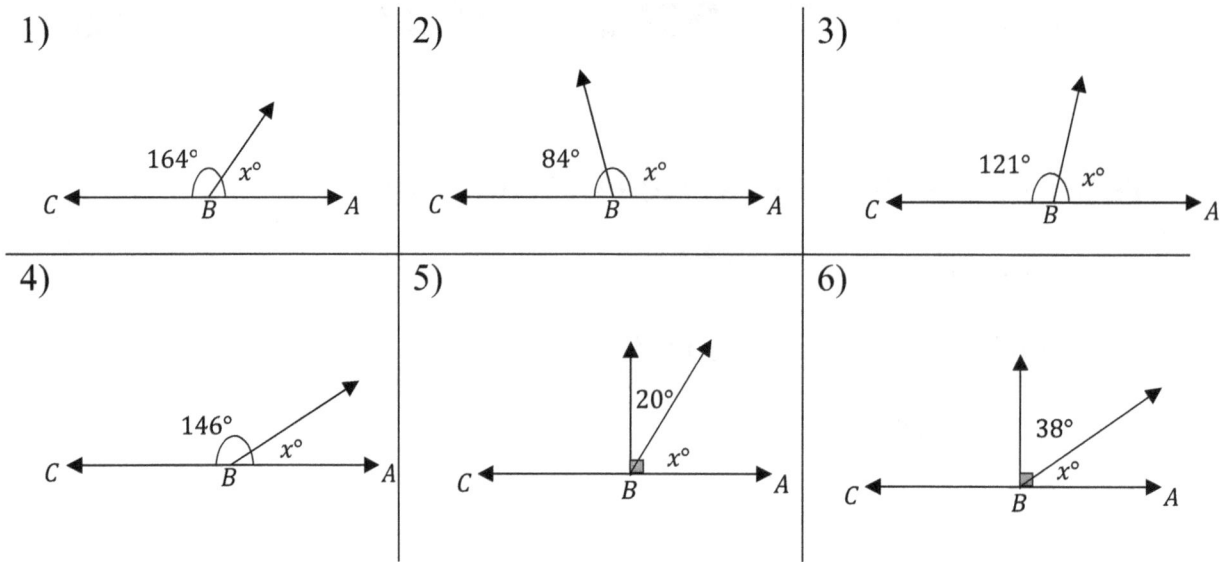

✎ **Calculate.**

7) Two supplement angles have equal measures. What is the measure of each angle? _____

8) The measure of an angle is seven fifth the measure of its supplement. What is the measure of the angle? _____

9) Two angles are complementary and the measure of one angle is 24 less than the other. What is the measure of the smaller angle? _____

10) Two angles are complementary. The measure of one angle is one fifth the measure of the other. What is the measure of the bigger angle? _____

11) Two supplementary angles are given. The measure of one angle is 40° less than the measure of the other. What does the smaller angle measure? _____

WWW.MathNotion.Com

HiSET Subject Test – Mathematics

Pythagorean Relationship

✎ Do the following lengths form a right triangle?

1)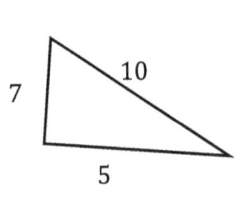

2)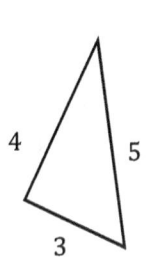

3)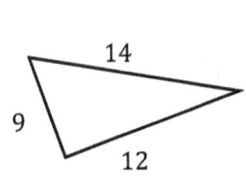

4)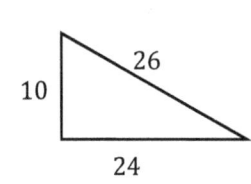

5)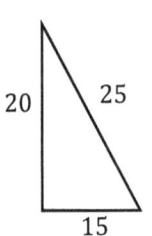

6)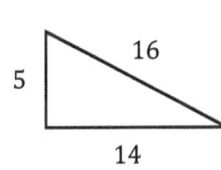

7)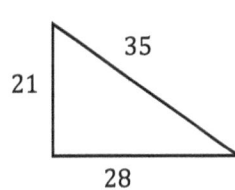

8)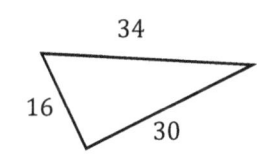

✎ Find the missing side?

9)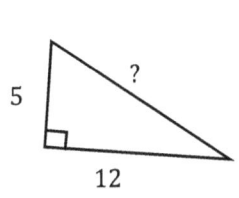

10)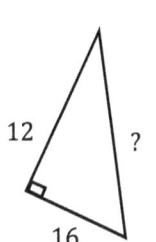

11)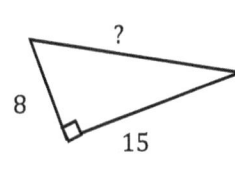

12)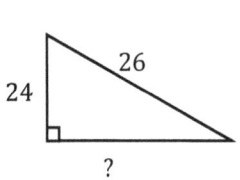

13)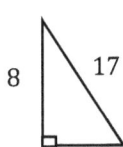

14)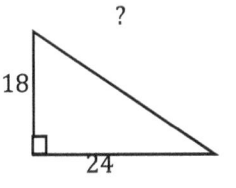

15)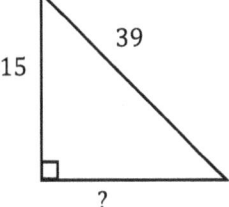

16)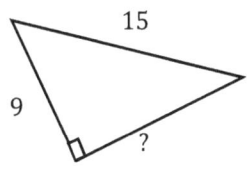

WWW.MathNotion.Com   125

HiSET Subject Test – Mathematics

Triangles

✎ Find the measure of the unknown angle in each triangle.

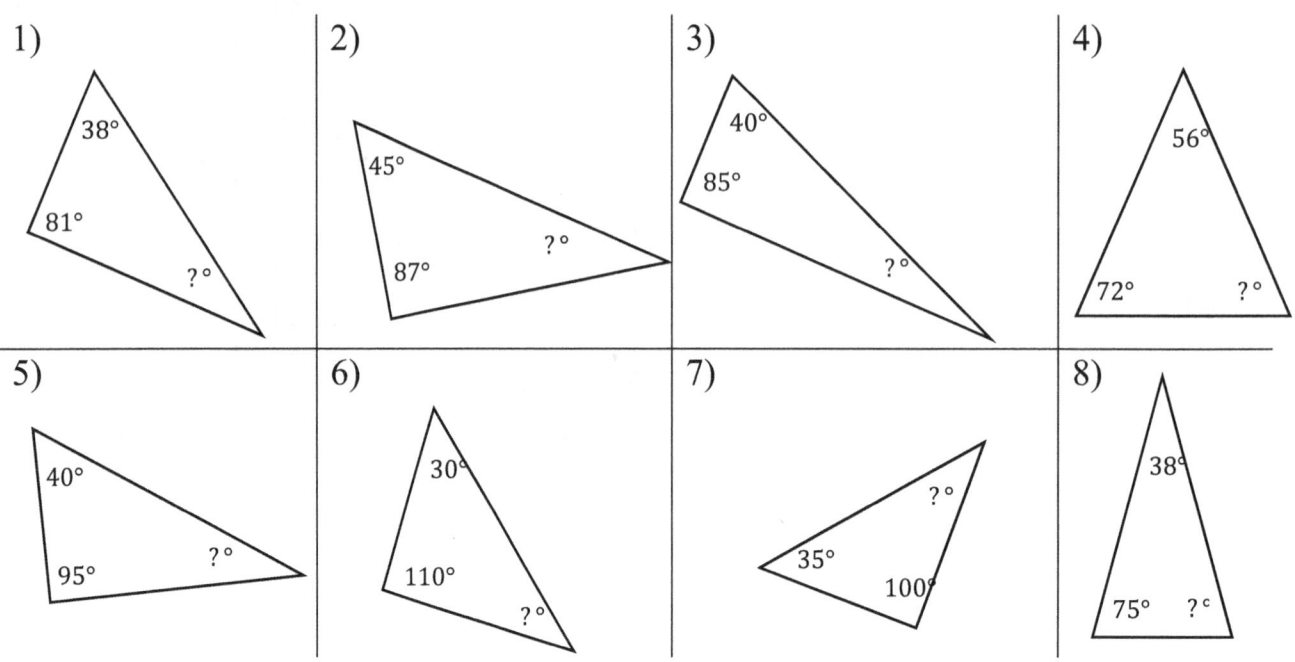

✎ Find area of each triangle.

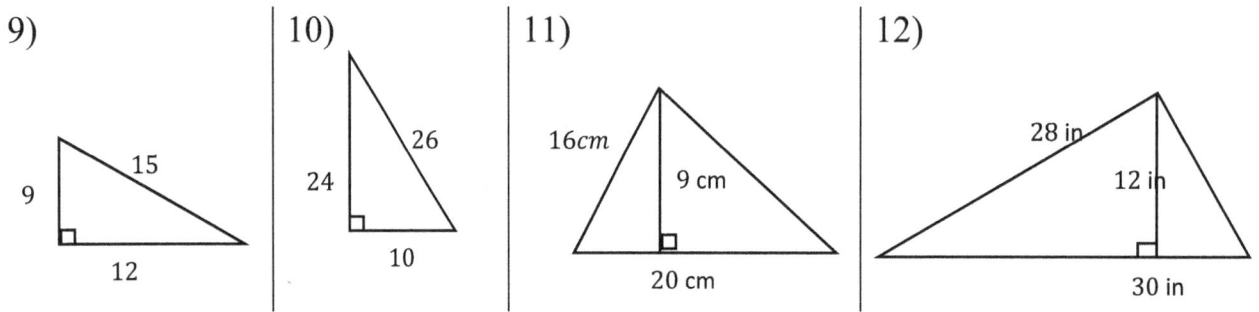

WWW.MathNotion.Com   126

HiSET Subject Test – Mathematics

Polygons

✏️ **Find the perimeter of each shape.**

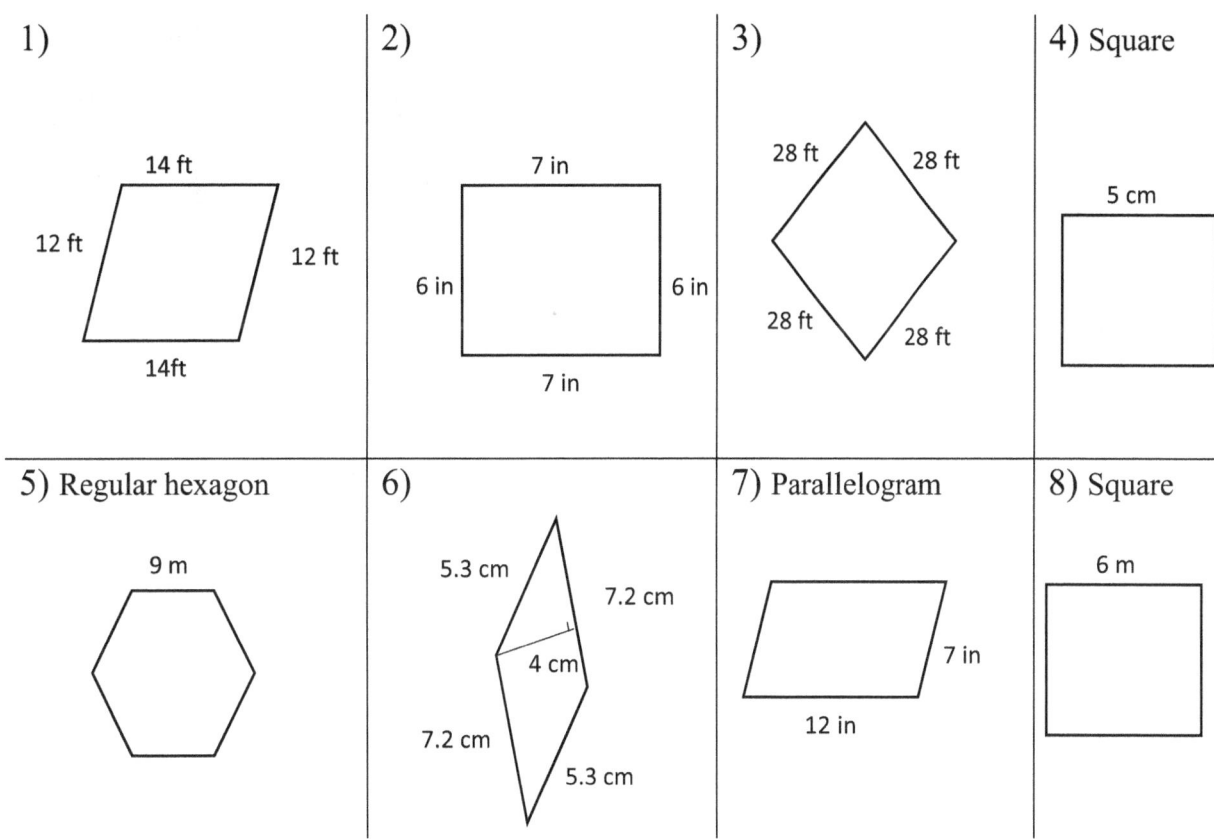

1)
14 ft
12 ft   12 ft
14ft

2)
7 in
6 in   6 in
7 in

3)
28 ft   28 ft
28 ft   28 ft

4) Square
5 cm

5) Regular hexagon
9 m

6)
5.3 cm
7.2 cm
4 cm
7.2 cm
5.3 cm

7) Parallelogram
7 in
12 in

8) Square
6 m

✏️ **Find the area of each shape.**

9) Parallelogram
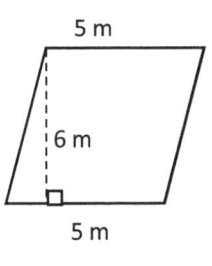
5 m
6 m
5 m

10) Rectangle
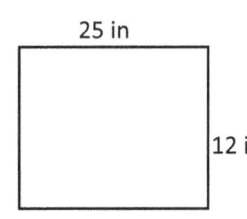
25 in
12 in

11) Rectangle
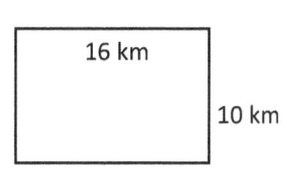
16 km
10 km

12) Square
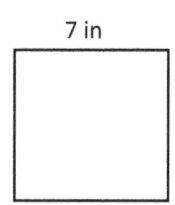
7 in

HiSET Subject Test – Mathematics

Trapezoids

✏️ **Find the area of each trapezoid.**

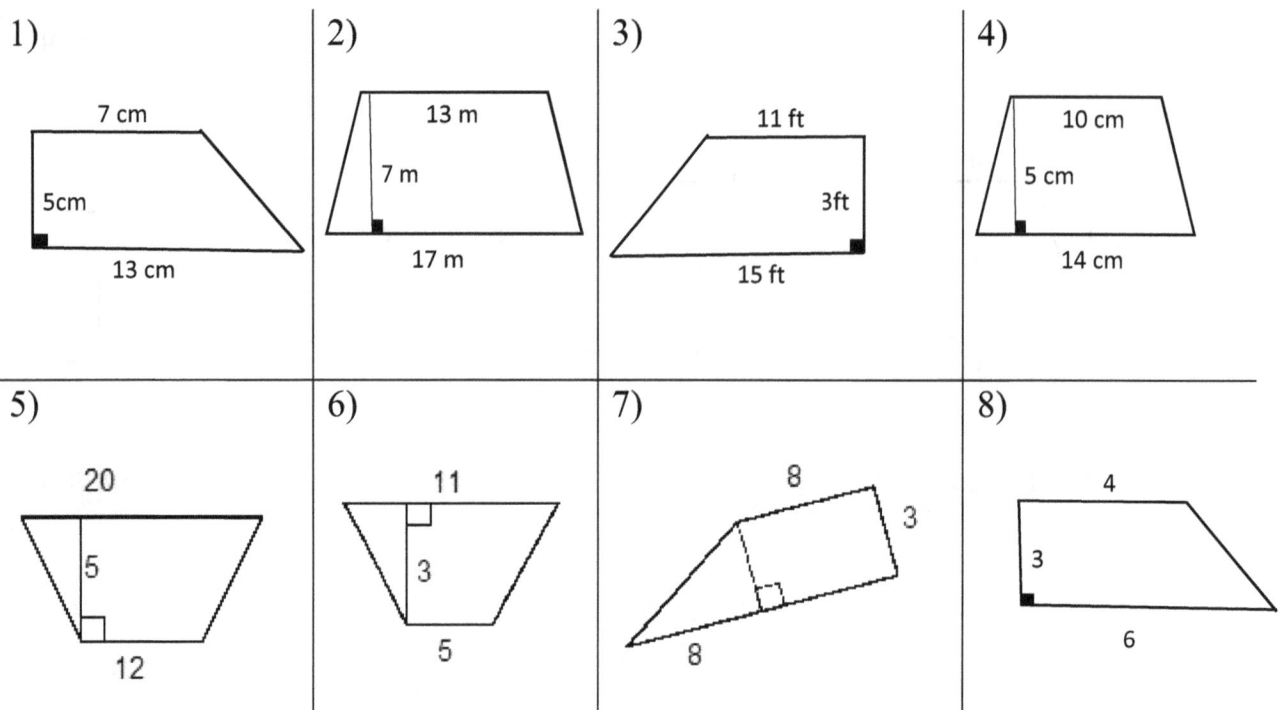

✏️ **Calculate.**

1) A trapezoid has an area of 45 cm² and its height is 5 cm and one base is 5 cm. What is the other base length? _____

2) If a trapezoid has an area of 99 ft² and the lengths of the bases are 8 ft and 10 ft, find the height? _____

3) If a trapezoid has an area of 126 m² and its height is 14 m and one base is 6 m, find the other base length? _____

4) The area of a trapezoid is 440 ft² and its height is 22 ft. If one base of the trapezoid is 15 ft, what is the other base length? _____

WWW.MathNotion.Com

HiSET Subject Test – Mathematics

Circles

✏️ **Find the area of each circle.** ($\pi = 3.14$)

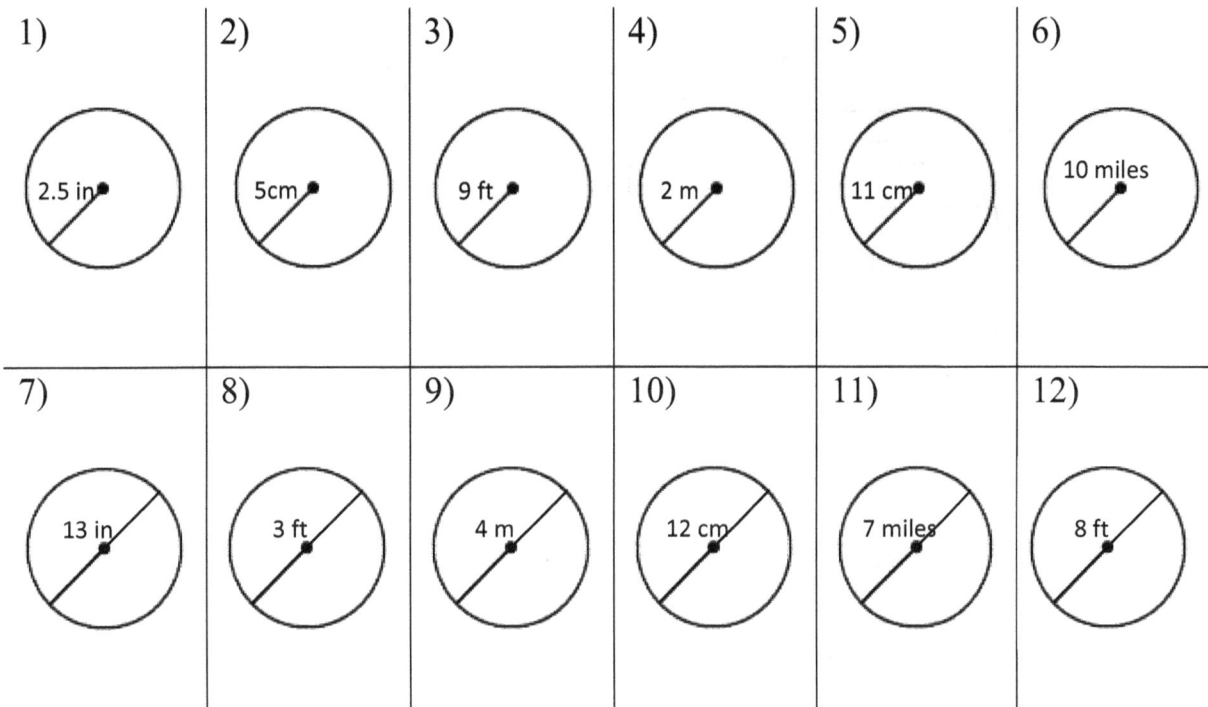

✏️ **Complete the table below.** ($\pi = 3.14$)

Circle No.	Radius	Diameter	Circumference	Area
1	1 in	2 in	6.28 in	3.14 in²
2		10 m		
3				28.26 ft²
4			47.1 mi	
5		11 km		
6	7 cm			
7		12 ft		
8				314 m²
9			56.52 in	
10	4.5 ft			

HiSET Subject Test – Mathematics

Cubes

✎ Find the volume of each cube.

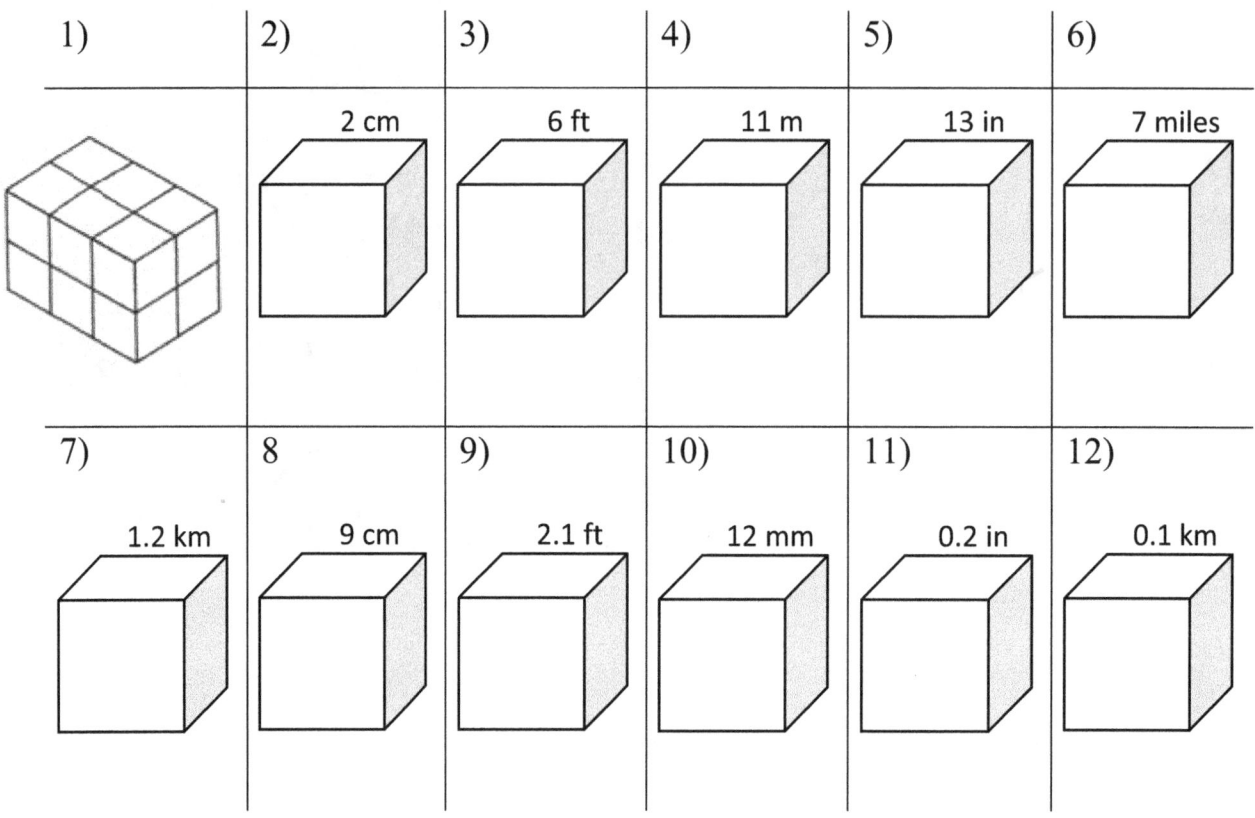

✎ Find the surface area of each cube.

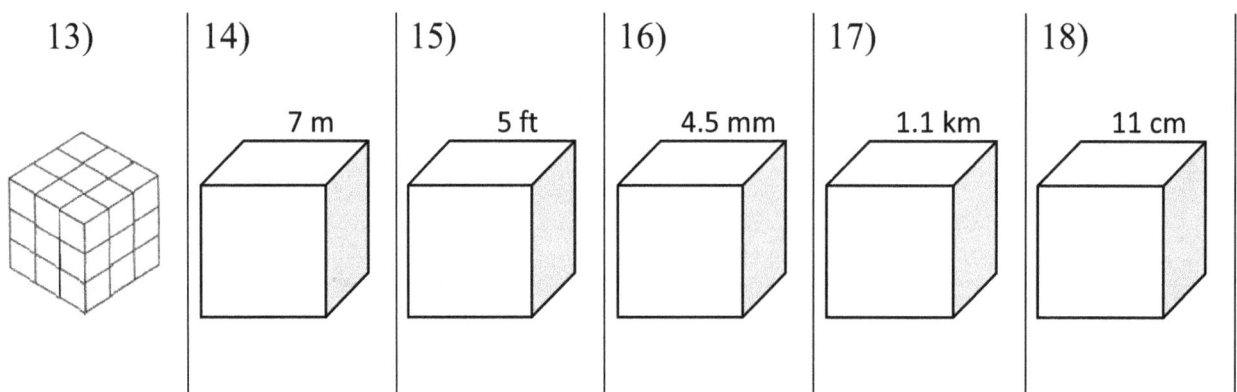

WWW.MathNotion.Com   130

HiSET Subject Test – Mathematics

Rectangular Prism

✏️ Find the volume of each Rectangular Prism.

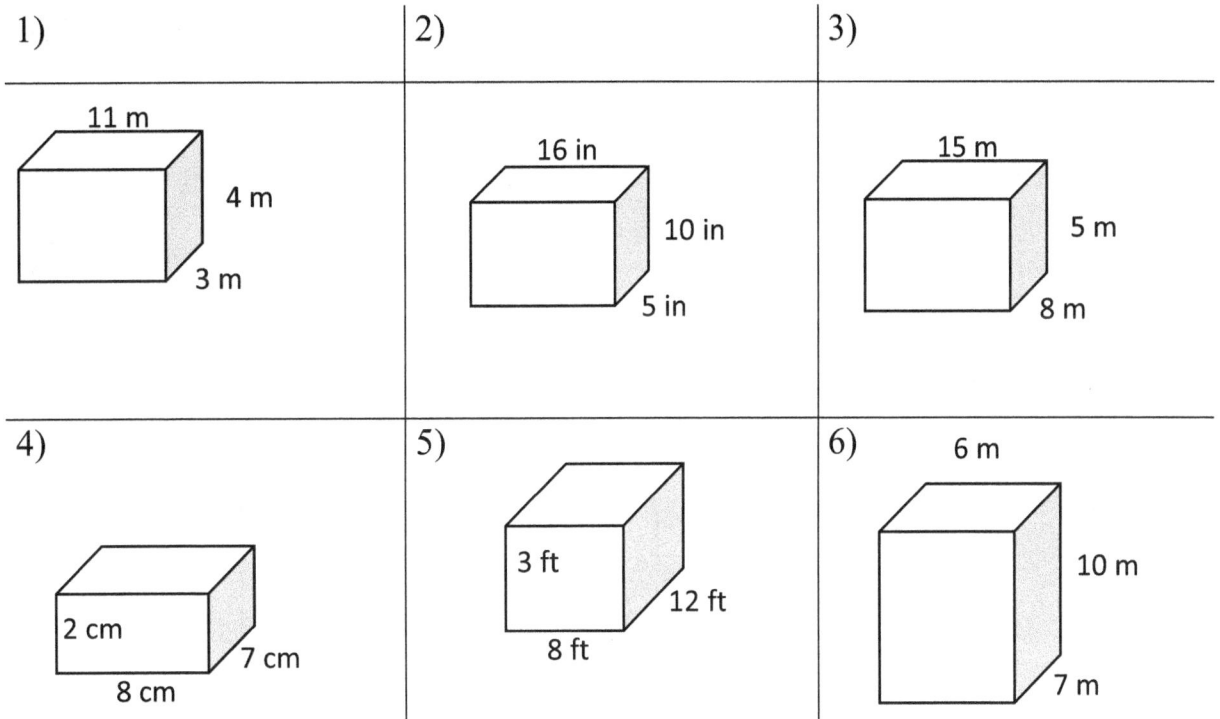

✏️ Find the surface area of each Rectangular Prism.

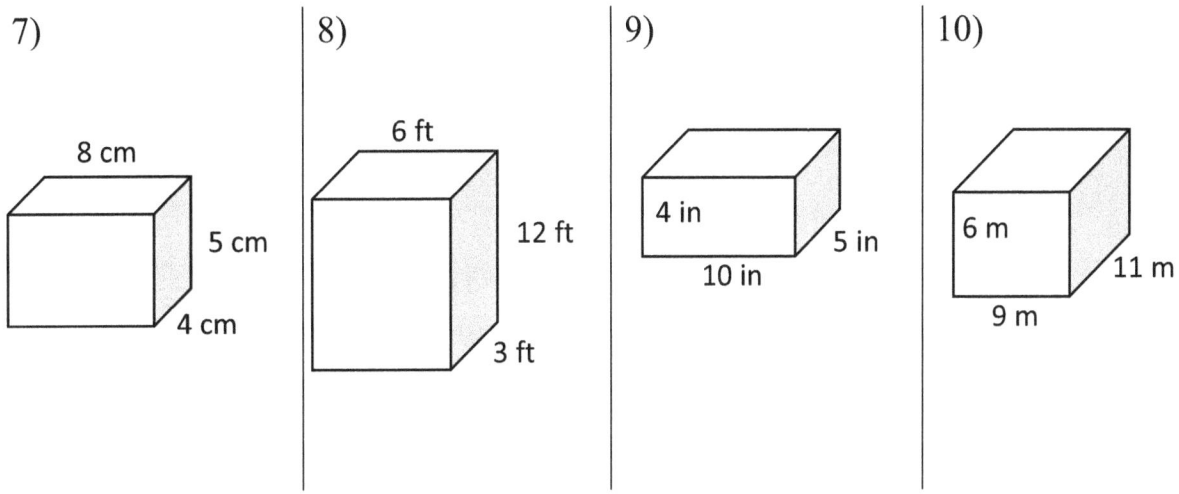

WWW.MathNotion.Com

HiSET Subject Test – Mathematics

Cylinder

✏️ **Find the volume of each Cylinder. Round your answer to the nearest tenth.** ($\pi = 3.14$)

1)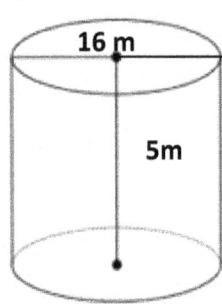

2)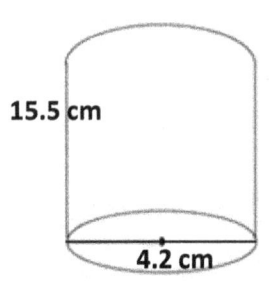

3)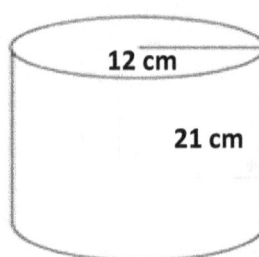

4)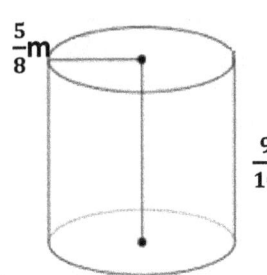

5)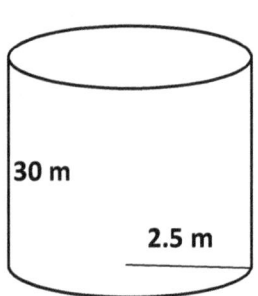

6)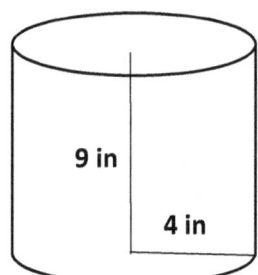

✏️ **Find the surface area of each Cylinder.** ($\pi = 3.14$)

7)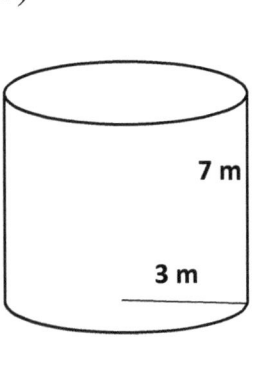

8)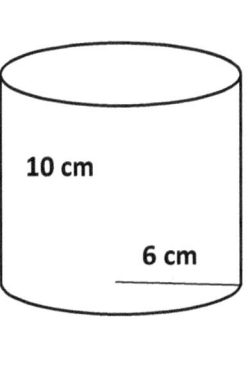

9)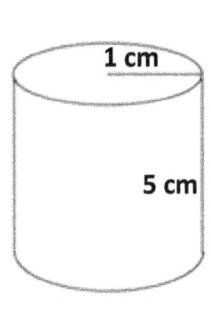

10)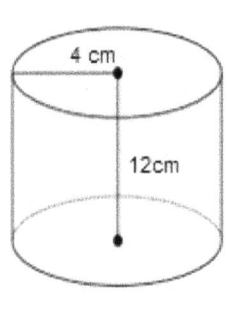

WWW.MathNotion.Com

HiSET Subject Test – Mathematics

Pyramids and Cone

✎ Find the volume of each Pyramid and Cone. ($\pi = 3.14$)

1)
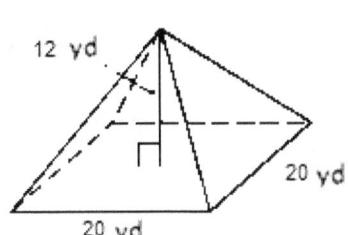

2)
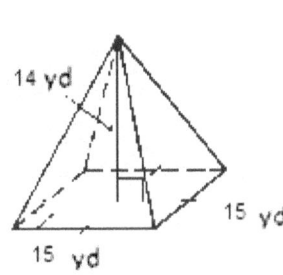

3)
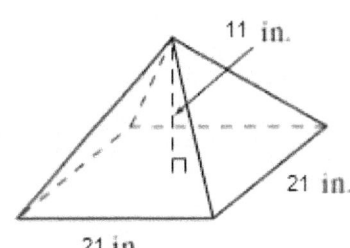

4)
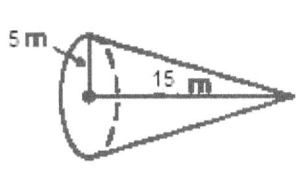

5)
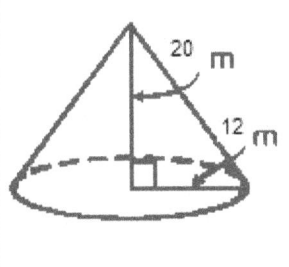

6)
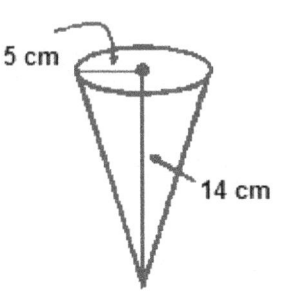

✎ Find the surface area of each Pyramid and Cone. ($\pi = 3.14$)

7)

8)

9)

10)

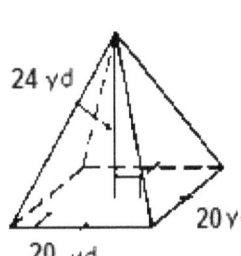
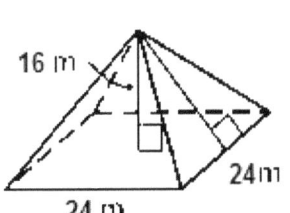
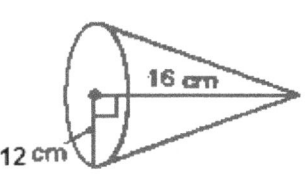

WWW.MathNotion.Com

HiSET Subject Test – Mathematics

Answers of Worksheets

Angles

1) 16°   4) 34°   7) 90°   10) 75°
2) 96°   5) 70°   8) 75°   11) 70°
3) 59°   6) 52°   9) 33°

Pythagorean Relationship

1) No   5) Yes   9) 13   13) 15
2) Yes   6) No   10) 20   14) 30
3) No   7) Yes   11) 17   15) 36
4) Yes   8) Yes   12) 10   16) 12

Triangles

1) 60°   5) 45°   9) 54 square unites
2) 48°   6) 40°   10) 120 square unites
3) 55°   7) 45°   11) 90 square unites
4) 52°   8) 67°   12) 180 square unites

Polygons

1) 52 ft   5) 54 m   9) 30 m^2
2) 26 in   6) 25 cm   10) 300 in^2
3) 112 ft   7) 38 in   11) 160 km^2
4) 20 cm   8) 24 m   12) 49 in^2

Trapezoids

1) 50 cm^2 4) 60 cm^2 7) 36
2) 105 m^2 5) 80   8) 15
3) 39 ft^2 6) 24

Calculate

1) 13 cm   2) 11 ft   3) 12 m   4) 25 ft

Circles

1) 19.63 in^2 5) 379.94 cm^2 9) 12.56 m^2
2) 78.5 cm^2 6) 314 $miles^2$ 10) 113.04 cm^2
3) 254.34 ft^2 7) 132.67 in^2 11) 38.47 $miles^2$
4) 12.56 m^2 8) 7.07 ft^2 12) 50.24 ft^2

HiSET Subject Test – Mathematics

Circle No.	Radius	Diameter	Circumference	Area
1	1 in	2 in	6.28 in	3.14 in^2
2	5 m	10 m	31.4 m	78.5 m^2
3	3 ft	6 ft	18.84 ft	28.26 ft^2
4	7.5 miles	15 mi	47.1 mi	176.63 mi^2
5	5.5 km	11 km	34.54 km	94.99 km^2
6	7 cm	14 cm	43.96 cm	153.86 cm^2
7	6 ft	12 ft	37.68 feet	113.04 ft^2
8	10 m	20 m	62.8 m	314 m^2
9	9 in	18 in	56.52 in	254.34 in^2
10	4.5 ft	9 ft	28.26 ft	63.585 ft^2

Cubes

1) 12
2) 8 cm^3
3) 216 ft^3
4) 1,331 m^3
5) 2,197 in^3
6) 343 $miles^3$
7) 1.728 km^3
8) 729 cm^3
9) 9.261 ft^3
10) 1,728 mm^3
11) 0.008 in^3
12) 0.001 km^3
13) 27
14) 294 m^2
15) 150 ft^2
16) 121.5 mm^2
17) 7.26 km^2
18) 726 cm^2

Rectangular Prism

1) 132 m^3
2) 800 in^3
3) 600 m^3
4) 112 cm^3
5) 288 ft^3
6) 420 m^3
7) 184 cm^2
8) 252 ft^2
9) 220 in^2
10) 438 m^2

Cylinder

1) 1,004.8 m^3
2) 214.6 cm^3
3) 9,495.4 cm^3
4) 1.1 m^3
5) 588.8 m^3
6) 452.2 in^3
7) 188.4 m^2
8) 602.9 cm^2
9) 37.7 cm^2
10) 401.9 m^2

Pyramids and Cone

1) 1,600 yd^3
2) 1,050 yd^3
3) 1,617 in^3
4) 392.5 m^3
5) 3,014.4 m^3
6) 366.33 cm^3
7) 1,440 yd^2
8) 1,536 m^2
9) 678.24 in^2
10) 1,205.76 cm^2

WWW.MathNotion.Com

HiSET Subject Test – Mathematics

HiSET Subject Test – Mathematics

Chapter 11:
Statistics and Probability

Topics that you'll practice in this chapter:

- ✓ Mean and Median
- ✓ Mode and Range
- ✓ Histograms
- ✓ Stem–and–Leaf Plot
- ✓ Pie Graph
- ✓ Probability Problems

Mathematics is no more computation than typing is literature.

− John Allen Paulos

HiSET Subject Test – Mathematics

Mean and Median

✎ **Find Mean and Median of the Given Data.**

1) 8, 7, 14, 4, 8

2) 14, 8, 25, 19, 16, 33, 11

3) 23, 18, 15, 12, 17

4) 34, 14, 10, 15, 6, 11

5) 10, 19, 6, 8, 32, 20, 17

6) 17, 26, 39, 69, 20, 6

7) 40, 38, 18, 11, 9, 2, 7, 32, 41

8) 24, 21, 31, 12, 33, 32, 22

9) 16, 14, 20, 41, 15, 20, 38, 4

10) 20, 20, 30, 18, 6, 28, 12, 46

11) 12, 7, 10, 11, 16, 22

12) 10, 29, 27, 12, 2, 15, 10, 3

✎ **Calculate.**

13) In a javelin throw competition, five athletics score 56, 34, 62, 23 and 19 meters. What are their Mean and Median? _____

14) Eva went to shop and bought 8 apples, 14 peaches, 6 bananas, 4 pineapples and 12 melons. What are the Mean and Median of her purchase? _____

15) Bob has 17 black pen, 19 red pen, 14 green pens, 20 blue pens and 5 boxes of yellow pens. If the Mean and Median are 19 respectively, what is the number of yellow pens in each box? _____

WWW.MathNotion.Com

HiSET Subject Test – Mathematics

Mode and Range

✎ **Find Mode and Rage of the Given Data.**

1) 4, 3, 7, 3, 3, 4
    Mode: _____   Range: _____

2) 18, 18, 24, 26, 18, 8, 14, 22
    Mode: _____   Range: _____

3) 8, 8, 8, 16, 19, 22, 20, 9, 13
    Mode: _____   Range: _____

4) 24, 24, 14, 28, 20, 18, 20, 24
    Mode: _____   Range: _____

5) 6, 21, 27, 24, 27, 27
    Mode: _____   Range: _____

6) 21, 8, 8, 7, 8, 12, 10, 22, 18, 13
    Mode: _____   Range: _____

7) 7, 4, 4, 6, 13, 13, 13, 0, 2, 2
    Mode: _____   Range: _____

8) 5, 8, 5, 14, 12, 14, 3, 5, 18
    Mode: _____   Range: _____

9) 7, 7, 7, 12, 7, 3, 8, 16, 3, 17
    Mode: _____   Range: _____

10) 15, 15, 19, 16, 4, 16, 10, 15
    Mode: _____   Range: _____

11) 6, 6, 5, 6, 42, 13, 19, 2
    Mode: _____   Range: _____

12) 8, 8, 9, 8, 9, 4, 34, 22
    Mode: _____   Range: _____

✎ **Calculate.**

13) A stationery sold 12 pencils, 56 red pens, 24 blue pens, 20 notebooks, 12 erasers, 21 rulers and 11 color pencils. What are the Mode and Range for the stationery sells?

    Mode: _____   Range: _____

14) In an English test, eight students score 10, 15, 15, 18 18, 16, 15 and 15. What are their Mode and Range? _____

15) What is the range of the first 6 even numbers greater than 8?
    _____

WWW.MathNotion.Com

HiSET Subject Test – Mathematics

Times Series

✎ Use the following Graph to complete the table.

Day	Distance (km)
1	
2	

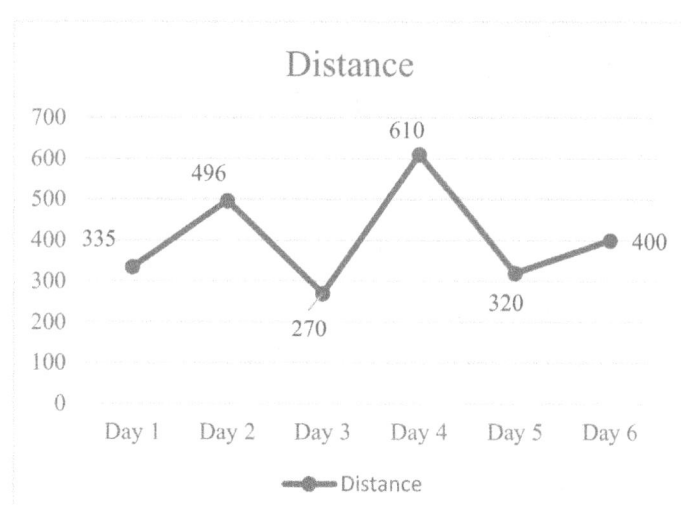

The following table shows the number of births in the US from 2007 to 2012 (in millions).

Year	Number of births (in millions)
2007	4.15
2008	3.70
2009	3.45
2010	3.20
2011	1.75
2012	2.98

Draw a Time Series for the table.

HiSET Subject Test – Mathematics

Stem–and–Leaf Plot

✍ **Make stem ad leaf plots for the given data.**

1) 24, 26, 29, 20, 53, 27, 51, 55, 36, 21, 37, 30   Stem | Leaf plot

2) 11, 59, 66, 14, 18, 19, 59, 65, 69, 61, 68, 65   Stem | Leaf plot

3) 121, 55, 66, 54, 112, 128, 63, 125, 59, 123, 68, 119   Stem | Leaf plot

4) 51, 32, 100, 56, 84, 36, 107, 56, 85, 39, 56, 106, 89   Stem | Leaf plot

5) 33, 89, 19, 87, 81, 16, 11, 30, 86, 35, 17, 35, 13   Stem | Leaf plot

6) 60, 92, 22, 25, 67, 93, 95, 62, 21, 64, 98, 29   Stem | Leaf plot

WWW.MathNotion.Com

Pie Graph

The circle graph below shows all Robert's expenses for last month. Robert spent $140 on his hobbies last month.

Answer following questions based on the Pie graph.

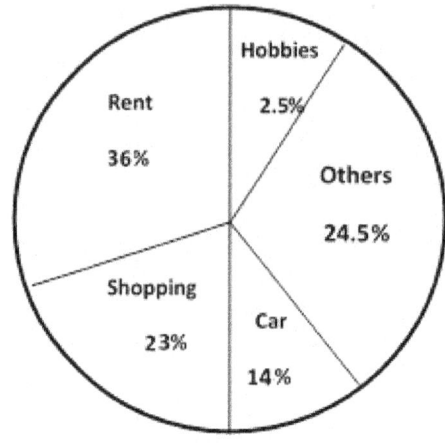

1) How much was Robert's total expenses last month? _____

2) How much did Robert spend on his car last month? _____

3) How much did Robert spend for shopping last month? _____

4) How much did Robert spend on his rent last month? _____

5) What fraction is Robert's expenses for his rent and car out of his total expenses last month? _____

HiSET Subject Test – Mathematics

Probability Problems

✎ **Calculate.**

1) A number is chosen at random from 1 to 10. Find the probability of selecting number 6 or smaller numbers. _____

2) Bag A contains 18 red marbles and 6 green marbles. Bag B contains 16 black marbles and 8 orange marbles. What is the probability of selecting a green marble at random from bag A? What is the probability of selecting a black marble at random from Bag B? _____

3) A number is chosen at random from 1 to 20. What is the probability of selecting multiples of 4? _____

4) A card is chosen from a well-shuffled deck of 52 cards. What is the probability that the card will be a queen? _____

5) A number is chosen at random from 1 to 15. What is the probability of selecting a multiple of 3 or 5? _____

A spinner numbered 1–8, is spun once. What is the probability of spinning …?

6) an Odd number? _____   7) a multiple of 2? _____

8) a multiple of 5? _____   9) number 10? _____

HiSET Subject Test – Mathematics

Answers of Worksheets

Mean and Median

1) Mean: 8.2, Median: 8
2) Mean: 18, Median: 16
3) Mean: 17, Median: 17
4) Mean: 15, Median: 12.5
5) Mean: 16, Median: 17
6) Mean: 29.5, Median: 23
7) Mean: 22, Median: 18
8) Mean: 25, Median: 24
9) Mean: 21, Median: 18
10) Mean: 22.5, Median: 20
11) Mean: 13, Median: 11.5
12) Mean: 13.5, Median: 11
13) Mean: 38.8, Median: 34
14) Mean: 8.8, Median: 8
15) 5

Mode and Range

1) Mode: 3, Range: 4
2) Mode: 18, Range: 18
3) Mode: 8, Range: 14
4) Mode: 24, Range: 14
5) Mode: 27, Range: 21
6) Mode: 8, Range: 15
7) Mode: 13, Range: 13
8) Mode: 5, Range: 15
9) Mode: 7, Range: 14
10) Mode: 15, Range: 15
11) Mode: 6, Range: 40
12) Mode: 8, Range: 30
13) Mode: 12, Range: 45
14) Mode: 15, Range: 8
15) 10

Time series

Day	Distance (km)
1	335
2	496
3	270
4	610
5	320
6	400

Stem–And–Leaf Plot

1)

Stem	leaf
2	0 1 4 6 7 9
3	0 6 7
5	1 3 5

2)

Stem	leaf
1	1 4 8 9
5	9 9
6	1 5 5 6 8 9

3)

Stem	leaf
5	4 5 9
6	3 6 8
11	2 9
12	1 3 5 8

WWW.MathNotion.Com

HiSET Subject Test – Mathematics

4)

Stem	leaf
3	2 6 9
5	1 6 6 6
8	4 5 9
10	0 6 7

5)

Stem	leaf
1	1 3 6 7 9
3	0 3 5 5
8	1 6 7 9

6)

Stem	leaf
2	2 1 5 9
6	0 2 4 7
9	2 3 5 8

Pie Graph

1) $5,600

2) $784

3) $1,288

4) $2,016

5) $\frac{1}{2}$

Probability Problems

1) $\frac{3}{5}$

2) $\frac{1}{4}, \frac{2}{3}$

3) $\frac{1}{4}$

4) $\frac{1}{13}$

5) $\frac{7}{15}$

6) $\frac{1}{2}$

7) $\frac{1}{2}$

8) $\frac{1}{8}$

9) 0

HiSET Subject Test – Mathematics

HiSET Subject Test – Mathematics

Chapter 12 : HiSET Test Review

The High School Equivalency Test (HiSET), commonly known as HiSET, is a standardized test and was released in the year 2014. This test was created by the ITP (Iowa Testing Programs) and ETS (Educational Testing Service). The HiSET is equal to the HiSET test. Currently, there are twelve states that offer the HiSET®: California, Iowa, Louisiana, Maine, Massachusetts, Missouri, Montana, Nevada, New Hampshire, New Jersey, Tennessee, and Wyoming.

HiSET test takers can choose to take the test using a computer, or with pencil and paper.

The HiSET is made up of five distinct sections:
- ❖ Social Studies,
- ❖ Language Arts Reading
- ❖ Language Arts Writing
- ❖ Science
- ❖ Mathematics

The HiSET Mathematics test is a 90-minute, single-section test that covers basic mathematics topics, quantitative problem-solving and algebraic questions. There are approximately 50 Multiple-choice questions on Mathematics section. Calculator is allowed in the Math section.

In this section, there are two complete HiSET Mathematics Tests. Take these tests to see what score you'll be able to receive on a real HiSET test.

HiSET Subject Test – Mathematics

Time to Test

Time to refine your skill with a practice examination

Take a REAL HiSET Mathematics test to simulate the test day experience. After you've finished, score your test using the answer key.

Before You Start

- You'll need a pencil, calculator, and a timer to take the test.
- It's okay to guess. You won't lose any points if you're wrong.
- After you've finished the test, review the answer key to see where you went wrong.

Calculators are permitted for the HiSET Mathematics Test.

Good Luck!

HiSET Mathematics Practice Tests Answer Sheet

Remove (photocopy) this answer sheet and use it to complete the practice test.

HiSET Mathematics Practice Test Answer Sheet

1	A B C D E	21	A B C D E	41	A B C D E
2	A B C D E	22	A B C D E	42	A B C D E
3	A B C D E	23	A B C D E	43	A B C D E
4	A B C D E	24	A B C D E	44	A B C D E
5	A B C D E	25	A B C D E	45	A B C D E
6	A B C D E	26	A B C D E	46	A B C D E
7	A B C D E	27	A B C D E	47	A B C D E
8	A B C D E	28	A B C D E	48	A B C D E
9	A B C D E	29	A B C D E	49	A B C D E
10	A B C D E	30	A B C D E	50	A B C D E
11	A B C D E	31	A B C D E		
12	A B C D E	32	A B C D E		
13	A B C D E	33	A B C D E		
14	A B C D E	34	A B C D E		
15	A B C D E	35	A B C D E		
16	A B C D E	36	A B C D E		
17	A B C D E	37	A B C D E		
18	A B C D E	38	A B C D E		
19	A B C D E	39	A B C D E		
20	A B C D E	40	A B C D E		

HiSET Subject Test – Mathematics

HiSET Subject Test – Mathematics

Practice Test 1

HiSET Mathematics

✓ 50 Questions

✓ Total time for this section: 90 Minutes

✓ You may use a calculator for this test.

Administered Month Year

HiSET Subject Test – Mathematics

1) Two fifth of 60 is equal to $\frac{4}{7}$ of what number?

   A. 24   C. 48   E. 52

   B. 35   D. 42

2) A bag contains 25 balls: two green, six black, seven blue, five red and five white. If 10 balls are removed from the bag at random, what is the probability that a white ball has been removed?

   A. $\frac{1}{25}$ C. $\frac{1}{10}$ E. $\frac{4}{25}$

   B. $\frac{1}{4}$ D. $\frac{1}{5}$

3) What is the value of 5^4?

   A. 125   C. 3,125   E. 1,225

   B. 625   D. 4,520

4) What is the median of these numbers? 26, 6, 39, 50, 21, 21, 12, 50, 43

   A. 24   C. 21   E. 26

   B. 23.5   D. 43

5) The marked price of a computer is E Euro. Its price decreased by 20% in March and later increased by 15 % in April. What is the final price of the computer in E Euro?

   A. 0.52 E   C. 0.08 E   E. 1.08 E

   B. 0.68E   D. 0.92 E

HiSET Subject Test – Mathematics

6) 44 is What percent of 40?

   A. 110 %   C. 40 %   E. 14%

   B. 140 %   D. 140 %

7) A rope weighs 520 grams per meter of length. What is the weight in kilograms of 7.5 meters of this rope? (1 kilograms = 1,000 grams)

   A. 39   C. 3.9   E. 39.0

   B. 0.39   D. 0.039

8) What is the value of x in the following figure?

   A. 50

   B. 150

   C. 140

   D. 160

   E. 30

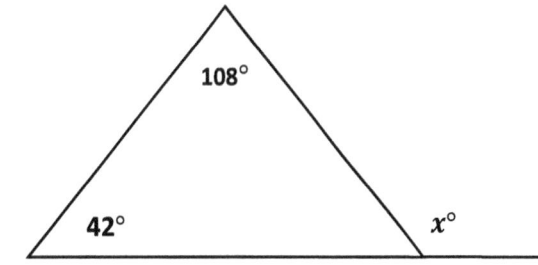

9) Which of the following could be the product of two consecutive prime numbers? (Select one or more answer choices)

   A. 125   C. 72   E. B, D

   B. 77   D. 36

10) A $80 shirt now selling for $32 is discounted by what percent?

   A. 40 %   C. 48 %   E. 12%

   B. 10 %   D. 60 %

HiSET Subject Test – Mathematics

11) The score of Zoe was one fourth of Emma and the score of Harper was twice that of Emma. If the score of Harper was 24, what is the score of Zoe?

    A. 3   C. 26   E. 2

    B. 12   D. 8

12) How many tiles of 6 cm² is needed to cover a floor of dimension 4 cm by 18 cm?

    A. 12   C. 72   E. 144

    B. 24   D. 16

13) $(x - 2y)(3x - 5y) = ?$

    A. $3x^2 - 10xy + 11y^2$ D. $3x^2 - 11xy + y^2$

    B. $3x^2 - 8xy + y^2$ E. $-6x^3 - 8y^3$

    C. $3x^2 - 11xy + 10y^2$

14) Ryan traveled 140 km in 7 hours and Riley traveled 240 km in 8 hours. What is the ratio of the average speed of Ryan to average speed of Riley?

    A. 3: 2   C. 3: 5   E. 2: 3

    B. 5: 3   D. 4: 3

15) An angle is equal to one fifth of its supplement. What is the measure of that angle?

    A. 30   C. 45   E. 25

    B. 15   D. 18

HiSET Subject Test – Mathematics

16) Abigail purchased a sofa for $286. The sofa is regularly priced at $440. What was the percent discount Abigail received on the sofa?

   A. 62 %   C. 65%   E. 24 %

   B. 38%   D. 35 %

17) Find the average of the following numbers: 32, 25, 34 and 19?

   A. 26.5   C. 42   E. 29

   B. 27.5   D. 22

18) When a number is subtracted from 54 and the difference is divided by that number, the result is 8. What is the value of the number?

   A. 9   C. 8   E. 6

   B. 12   D. 4

19) Right triangle ABC has two legs of lengths 15 cm (AB) and 20 cm (AC). What is the length of the third side (BC)?

   A. 36 cm   C. 25 cm   E. 40 cm

   B. 22 cm   D. 35 cm

20) If the area of trapezoid is 330 cm^2, what is the perimeter of the trapezoid?

   A. 70

   B. 85

   C. 64

   D. 75

   E. 80

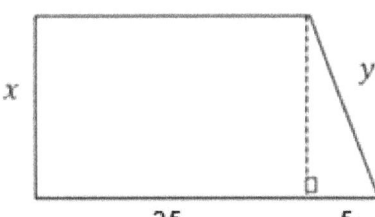

WWW.MathNotion.Com

HiSET Subject Test – Mathematics

21) A taxi driver earns $15 per hour work. If he works 12 hours a day, and he uses 5-liters Petrol in 2 hours with price $2.50 for 1-liter. How much money does he earn in one day?

    A. $105   C. $106.5   E. $75

    B. $180   D. $150.5

22) Solve for x:
$$4(x - 2) + 36 = 3(x + 1)$$

    A. 6   C. 10   E. -24

    B. -10 D. -25

23) The width of a box is one ninth of its length. The height of the box is half of its width. If the length of the box is 72 cm, what is the volume of the box?

    A. 288 cm^3 C. 576 cm^3 E. 2,340 cm^3

    B. 1,152 cm^3 D. 2,304 cm^3

24) The price of a sofa is decreased by 20% to $592. What was its original price?

    A. $1,960   C. $740   E. $118.4

    B. $2,960   D. $540

25) If 70% of a class are girls, and 40% of girls play tennis, what percent of the class play tennis?

    A. 47%   C. 74%   E. 36%

    B. 30%   D. 28%

HiSET Subject Test – Mathematics

26) The price of a car was $42,000 in 2014, $23,100 in 2015 and $12,705 in 2016. What is the rate of depreciation of the price of car per year?

A. 25 %

B. 10 %

C. 45 %

D. 35 %

E. 30%

27) Which of the following expressions is equivalent to $4xy(x^2 + y^2)$?

A. $4yx^2 + 4xy^2$

B. $4yx^3 - 4xy^2$

C. $4x^3 + 4y^3$

D. $4xy^3 + 4x^3y$

E. $4xy^3 - 4yx^3$

28) The average of five consecutive numbers is 42. What is the smallest number?

A. 43

B. 42

C. 41

D. 44

E. 40

29) The area of a circle is less than 81π. Which of the following can be the circumference of the circle? (Select one or more answer choices)

A. 22π

B. 18π

C. 25π

D. 16π

E. 20π

30) In three successive hours, a car travels 72 km, 63 km, and 45 km. In the next five hours, it travels with an average speed of 48 km per hour. Find the total distance the car traveled in 8 hours.

A. 420 km

B. 250 km

C. 320 km

D. 220 km

E. 180 km

HiSET Subject Test – Mathematics

31) A bank is offering 2.25% simple interest on a savings account. If you deposit $5,000, how much interest will you earn in eight years?

   A. $600

   B. $672

   C. $900

   D. $1,260

   E. $787.5

32) If $y = 2c^3 - 8cd + 5d^2$, what is y when $c = -2$, and $d = 2$?

   A. 16

   B. 42

   C. 46

   D. 52

   E. 36

33) The ratio of boys to girls in a school is 4: 3. If there are 420 students in a school, how many boys are in the school.

   A. 120

   B. 180

   C. 220

   D. 240

   E. 60

34) In the xy-plane, the point $(0, -1)$ and $(2, 7)$ are on the line A. Which of the following points could also be on the line A? (Select one or more answer choices)

   A. $(3, 9)$

   B. $(1, 5)$

   C. $(-1, -7)$

   D. $(0, -1)$

   E. $(4, 12)$

35) In 1999, the average worker's income increased $1,500 per year starting from $42,300 annual salary. Which equation represents income greater than average? (I = income, x = number of years after 1999)

   A. $I > 1,500 + 42,300$

   B. $I > 1,500 - 42,300$

   C. $I > -1,500 + 42,300$

   D. $I < 42,300x + 1,500$

   E. $I < -42,300x + 1,500$

HiSET Subject Test – Mathematics

36) If $f(x) = x^2 - 5$, What is the average rate of change of the function from $x = 2$ to $x = 5$?

   A. 5   C. -7 E. 4

   B. -5 D. 7

37) The average high of 15 constructions in a town is 140 m, and the average high of 10 towers in the same town is 160 m. What is the average high of all the 25 structures in that town?

   A. 55   C. 148   E. 152

   B. 72   D. 138

38) A chemical solution contains 8% alcohol. If there is 28 ml of alcohol, what is the volume of the solution?

   A. 1,350 ml   C. 1,400 ml   E. 350 ml

   B. 2,400 ml   D. 530 ml

39) The price of a laptop is decreased by 30% to $315. What is its original price?

   A. 350   C. 420   E. 550

   B. 255   D. 450

40) If 90 % of F is 15 % of M, then F is what percent of M?

   A. 6 %   C. 0.06%   E. 600 %

   B. 0.006 %   D. 60 %

HiSET Subject Test – Mathematics

41) What are the zeros of the function: $f(x) = 2x^3 - 4x^2 - 30x$?

A. $3, -5$ C. $3, 5$ E. $0, -3, -5$

B. $-3, -5$ D. $0, -3, 5$

42) A boat sails 160 miles south and then 120 miles east. How far is the boat from its start point?

A. 180 miles C. 280 miles E. 200 miles

B. 300 miles D. 250 miles

43) How many possible outfit combinations come from five shirts, two slacks, and 6 times?

A. 30 C. 12! E. 60!

B. 60 D. 30!

44) The surface area of a cylinder is $72\pi\ cm^2$. If its height is 5 cm, what is the radius of the cylinder?

A. 6 cm C. 14 cm E. 9 cm

B. 7 cm D. 4 cm

45) A shirt costing $900 is discounted 22%. After a month, the shirt is discounted another 7%. Which of the following expressions can be used to find the selling price of the shirt?

A. $(900) - 900(0.22)$ D. $(900)(0.22) - (900)(0.07)$

B. $(900)(0.22)(0.07)$ E. $(900)(0.22) - (900)(0.7)$

C. $(900)(0.78)(0.93)$

HiSET Subject Test – Mathematics

46) How long does a 228–miles trip take moving at 40 miles per hour(mph)?

   A. 5 hours and 42 minutes

   B. 5 hours and 15 minutes

   C. 6 hours and 20 minutes

   D. 5 hours

   E. 6 hours

47) Which of the following lists shows the fractions in order from least to greatest?

   A. $\frac{1}{4}, \frac{5}{16}, \frac{4}{7}$

   B. $\frac{4}{7}, \frac{5}{16}, \frac{1}{4}$

   C. $\frac{5}{16}, \frac{1}{4}, \frac{4}{7}$

   D. $\frac{4}{7}, \frac{1}{4}, \frac{5}{16}$

   E. $\frac{1}{4}, \frac{4}{7}, \frac{5}{16}$

48) Multiply and write the product in scientific notation:

$$(4.2 \times 10^9) \times (3.5 \times 10^{-11})$$

   A. 147×10^2

   B. 147×10^{-2}

   C. 147×10^{-99}

   D. 1,470

   E. 0.147

49) What is the value of y in the following system of equation?

$$2x + 3y = -8$$

$$3x - 4y = 5$$

   A. −3

   B. −2

   C. −4

   D. 1

   E. 6

HiSET Subject Test – Mathematics

50) If the height of a right pyramid is triple of its side, and its base is a square with side 5 cm. What is its volume?

A. 75 cm^3

B. 25 cm^3

C. 125 cm^3

D. 150 cm^3

E. 372 cm^3

HiSET Subject Test – Mathematics

Practice Test 2

HiSET Mathematics

✓ **50 Questions**

✓ **Total time for this section: 90 Minutes**

✓ You may use a calculator for this test.

Administered *Month Year*

HiSET Subject Test – Mathematics

1) The mean of 90 test scores was calculated as 65. But it turned out that one of the scores was misread as 73 but it was 37. What is the correct mean of the test scores?

   A. 73

   B. 82

   C. 64.6

   D. 46.6

   E. 67.5

2) Which of the following graphs represents the compound inequality $-2 \leq 3x - 5 < 19$?

   A.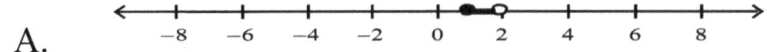

   B.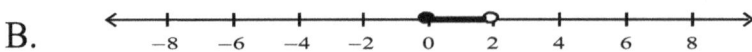

   C.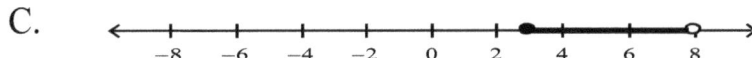

   D.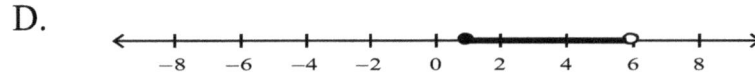

   E.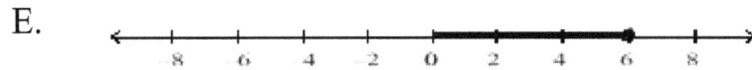

3) Two dice are thrown simultaneously, what is the probability of getting a sum of 5 or 8?

   A. $\frac{7}{12}$

   B. $\frac{7}{36}$

   C. $\frac{1}{36}$

   D. $\frac{1}{4}$

   E. $\frac{1}{5}$

4) Simplify the expression.

$$(6x^3 - 3x^2 - 5x^4) - (4x^2 - 2x^4 - 5x^3)$$

   A. $-(2x^4 + 11x^3 - 7x^2)$

   B. $(3x^4 + 7x^3 - 11x^2)$

   C. $-(3x^4 + 7x^3 - 11x^2)$

   D. $-(3x^4 - 11x^3 + 7x^2)$

   E. $(-3x^4 - 11x^3 - 7x^2)$

HiSET Subject Test – Mathematics

5) What is the perimeter of a square in centimeters that has an area of 812.25 cm²?

   A. 812.25 C. 406.1 E. 3,249

   B. 114 D. 57.25

6) If 60% of A is 12% of B, then B is what percent of A?

   A. 50% C. 0.05% E. 500%

   B. 80% D. 5%

7) If $f(x) = 2x^3 + 3x^2 - 2x + 9$ and $g(x) = 3$, what is the value of $(f \circ g)(x) = ?$

   A. −48 C. −75 E. 84

   B. 75 D. 48

8) Mr. Matthews saves $3,500 out of his annually family income of $56,000. What fractional part of his income does he save?

   A. $\frac{1}{16}$ C. $\frac{5}{16}$ E. 16

   B. $\frac{9}{16}$ D. $\frac{3}{16}$

9) What is the median of these numbers? 32, 16, 11, 41, 35, 24, 58

   A. 47 C. 41 E. 16

   B. 24 D. 32

10) A bank is offering 2.25% simple interest on a savings account. If you deposit $23,000, how much interest will you earn in six years?

   A. $1,035 C. $3,050 E. $3,450

   B. $3,105 D. $1,350

WWW.MathNotion.Com

HiSET Subject Test – Mathematics

11) What are the zeros of the function: $f(x) = 2x^2 + 4x - 48$?

   A. $-2, -6$ C. $1, -4$ E. $4, -4$

   B. $-6, 4$ D. $4, -6$

12) Last week 15,000 fans attended a football match. This week six times as many bought tickets, but one fifth of them cancelled their tickets. How many are attending this week?

   A. 90,000   C. 20,000   E. 12,000

   B. 108,000   D. 72,000

13) In two successive years, the population of a town is increased by 25% and 15%. What percent of the population is increased after two years?

   A. 56.2%   C. 3.75%   E. 156.2%

   B. 43.8%   D. 143.8%

14) Which of the following shows the numbers in descending order?

   $$\frac{1}{25}, 0.05, 7\%, \frac{1}{12}$$

   A. $7\%, 0.05, \frac{1}{12}, \frac{1}{25}$ D. $\frac{1}{12}, 0.05, 7\%, \frac{1}{25}$

   B. $7\%, 0.05, \frac{1}{25}, \frac{1}{12}$ E. $\frac{1}{12}, 7\%, 0.05, \frac{1}{25}$

   C. $\frac{1}{12}, \frac{1}{25}, 7\%, 0.05$

15) What is the volume of a box with the following dimensions?

   Height = 5 cm   Width = 6 cm   Length = 9 cm

   A. 270 cm³   C. 30 cm³   E. 135 cm³

   B. 45 cm³   D. 54 cm³

WWW.MathNotion.Com

HiSET Subject Test – Mathematics

16) What is the area of a square whose diagonal is 8?

   A. 16   C. 32   E. 64

   B. 14   D. 25

17) The average of 8 numbers is 25. The average of 6 of those numbers is 29. What is the average of the other two numbers?

   A. 26   C. 13   E. 19

   B. 18   D. 14

18) What is the value of y in the following system of equations?

$$4x + 5y = -12$$
$$4x + 2y = 6$$

   A. 3   C. 6   E. $\frac{9}{2}$

   B. $-\frac{9}{2}$ D. -6

19) If a tree casts a 63–inches shadow while a 11-inches yardstick casts a 9–inches shadow, what is the height of the tree?

   A. 54 in

   B. 154 in

   C. 77 in

   D. $\frac{63}{11}$ in

   E. 51 in

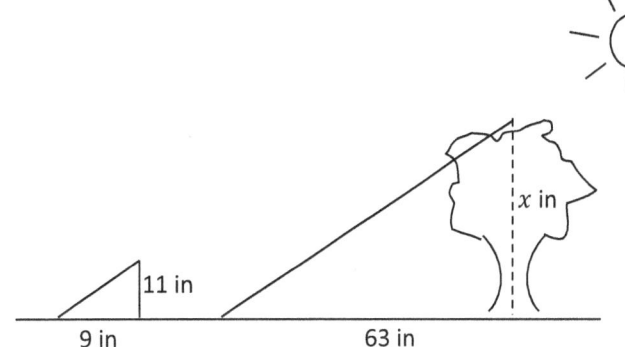

HiSET Subject Test – Mathematics

20) The perimeter of a rectangular yard is 56 meters. What is its length if its width is triple length?

   A. 7 meters   C. 21 meters   E. 28 meters

   B. 8 meters   D. 24 meters

21) In a stadium the ratio of home fans to visiting fans in a crowd is 9:4. Which of the following could be the total number of fans in the stadium? (Select one or more answer choices)

   A. 52,630   C. 34,620   E. 48,123

   B. 41,249   D. 72,936

22) What is the equivalent temperature of $149°F$ in Celsius? $C = \frac{5}{9}(F - 32)$

   A. 59   C. 65   E. 66

   B. 36   D. 56

23) Which of the following points lies on the line $5x - 2y = 1$? (Select one or more answer choices)

   A. (2, 1)   C. (1, 2)   E. (3, 0)

   B. (−1, 3)   D. (−2, 2)

24) Mr. Jefferson family are choosing a menu for their reception. They have 5 choices of appetizers, 9 choices of entrees, 4 choices of cake. How many different menu combinations are possible for them to choose?

   A. 45   C. 36   E. 180

   B. 190   D. 20

HiSET Subject Test – Mathematics

25) Which of the following is equal to the expression below? $(x - 3y)^2$

   A. $x^2 - 9y^2$

   B. $2x^2 - 9y^2$

   C. $x^2 - 6xy + 9y^2$

   D. $x^2 + 3xy + 9y^2$

   E. $x^2 + 6xy + 9y^2$

26) Anita's trick–or–treat bag contains 28 pieces of chocolate, 21 suckers, 20 pieces of gum, 11 pieces of licorice. If she randomly pulls a piece of candy from her bag, what is the probability of her pulling out a piece of gum?

   A. $\frac{1}{8}$

   B. $\frac{3}{20}$

   C. $\frac{1}{5}$

   D. $\frac{1}{4}$

   E. $\frac{3}{40}$

27) The average of six numbers is 70. If a seventh number that is greater than 77 is added, then, which of the following could be the new average? (Select one or more answer choices)

   A. 74

   B. 70

   C. 69

   D. 68

   E. 71

28) The perimeter of the trapezoid below is 52 cm. What is its area?

   A. 90 cm²

   B. 154 cm²

   C. 150 cm²

   D. 308 cm²

   E. 168 cm²

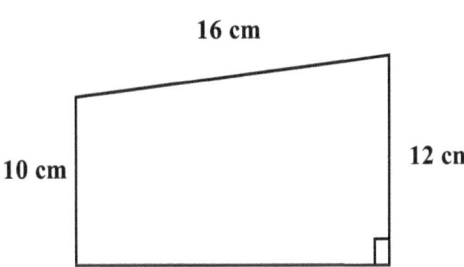

HiSET Subject Test – Mathematics

29) A card is drawn at random from a standard 52–card deck, what is the probability that the card is of spades or hearts? (The deck includes 13 of each suit clubs, diamonds, hearts, and spades).

A. $\frac{1}{4}$

B. $\frac{1}{2}$

C. $\frac{4}{13}$

D. $\frac{1}{8}$

E. $\frac{2}{13}$

30) What is the value of the expression $-5(2x + y) + (6 - x)^2$ when $x = 2$ and $y = -3$?

A. 22

B. −21

C. 21

D. −11

E. 11

31) The ratio of boys and girls in a class is 3:7. If there are 120 students in the class, how many more boys should be enrolled to make the ratio 1:1?

A. 48

B. 84

C. 36

D. −42

E. 46

32) What is the surface area of the cylinder below?

A. 290π in²

B. 384π in²

C. 250π in²

D. 160π in²

E. 80π in²

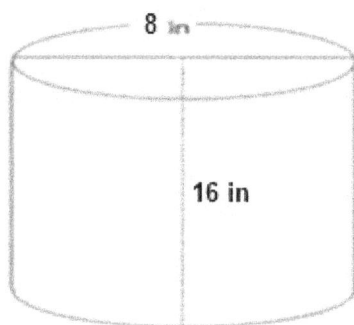

HiSET Subject Test – Mathematics

33) Simplify $3x^3y^4(-2xy^2)^2$.

A. $12x^4y^8$

B. $-6x^5y^8$

C. $12x^5y^8$

D. $12x^5y^{10}$

E. $6x^5y^8$

34) Daniel is 54 miles ahead of Noa and running at 4.5 miles per hour. Noa is running at the speed of 9 miles per hour. How long does it take Noa to catch Daniel?

A. 11 hours, and 30 minutes

B. 12 hours, and 30 minutes

C. 11 hours, 40 minutes

D. 12 hours

E. 11 hours

35) A football team had $56,000 to spend on supplies. The team spent $36,000 on new balls. New sport shoes cost $160 each. Which of the following inequalities represent the number of new shoes the team can purchase?

A. $160x + 36,000 \leq 56,000$

B. $160x + 20,000 \leq 56,000$

C. $160x + 56,000 \geq 36,000$

D. $56,000 + 160x \geq 20,000$

E. $36,000 + 160x \geq 56,000$

36) 75 students took an exam and 24 of them failed. What percent of the students passed the exam?

A. 63%

B. 12.5%

C. 8%

D. 68%

E. 32%

HiSET Subject Test – Mathematics

37) The length of a rectangle is 4 meters less than 7 times its width. The perimeter of the rectangle is 88 meters. What is the area of the rectangle in meters?

   A. 222 C. 228 E. 266

   B. 195 D. 288

38) If 80% of a number is 36, what is the number?

   A. 24 C. 44 E. 45

   B. 52 D. 54

39) The square of a number is $\frac{80}{125}$. What is the cube of that number?

   A. $\frac{32}{50}$ C. $\frac{64}{625}$ E. $\frac{1,024}{3,125}$

   B. $\frac{64}{125}$ D. $\frac{256}{625}$

40) The circle graph below shows all Mr. Wilson's expenses for last month. If he spent $640 on his car, how much did he spend for his rent?

   A. $960

   B. $3,200

   C. $1,690

   D. $96

   E. $0.96

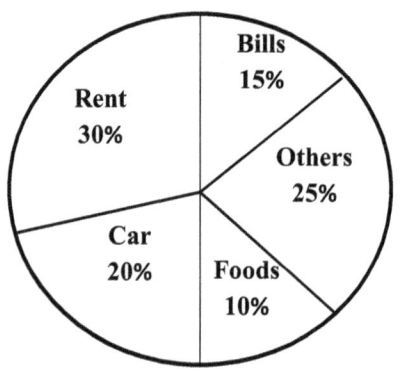

Mr. Wilson's monthly expenses

HiSET Subject Test – Mathematics

41) There are two equal tanks of water. If $\frac{3}{5}$ of a tank contains 420 liters of water, what is the capacity of the three tanks of water together?

    A. 1,400 L   C. 2,100 L   E. 700 L

    B. 84 L   D. 2,800 L

42) What is the positive value of x in the following equation?

$$|-4x + 8| = 36$$

    A. 11   C. 12   E. 8

    B. 7   D. 10

43) If 140% of a number is 126, then what is the 40% of that number?

    A. 63   C. 36   E. 54

    B. 45   D. 56

44) A swimming pool holds 6,300 cubic feet of water. The swimming pool is 30 feet long and 15 feet wide. How deep is the swimming pool?

    A. 1.6   C. 6.5   E. 450

    B. 14   D. 16

45) What is the value of x in the following equation?

$$\frac{5}{12}x + \frac{1}{6} = \frac{3}{4}$$

    A. 5   C. 7   E. $\frac{5}{7}$

    B. $\frac{1}{5}$ D. $\frac{7}{5}$

WWW.MathNotion.Com

HiSET Subject Test – Mathematics

46) Mrs. Thomson needs an 76% average in her writing class to pass. On her first 4 exams, he earned scores of 78%, 81%, 82%, and 89%. What is the minimum score Mrs. Thomson can earn on her fifth and final test to pass?

A. 50   C. 65   E. 85

B. 55   D. 75

47) Right triangle ABC has two legs of lengths 18 cm (AB) and 24 cm (AC). What is the length of the third side (BC)?

A. 35 cm   C. 50 cm   E. 30 cm

B. 60 cm   D. 40 cm

48) What is the slope of a line that is perpendicular to the line $7x - 2y = 8$?

A. $\frac{2}{7}$ C. $-\frac{7}{2}$ E. $\frac{1}{7}$

B. $-\frac{2}{7}$ D. $\frac{7}{2}$

49) Which set of ordered pairs models a function?

A. $\{(2,8), (4,-1), (2,1), (7,-2)\}$

B. $\{(6,5), (3,0), (3,12), (3,8)\}$

C. $\{(-5,-1), (1,-7), (-5,2), (1,9)\}$

D. $\{(8,4), (5,3), (2,2), (4,1)\}$

E. $\{(7,3), (-5,10), (7,8), (-7,19)\}$

50) $49 + 7 \times (-16) \div 14 - 9 = ?$

A. 42   C. 30   E. 32

B. 38   D. 34

HiSET Subject Test – Mathematics

Chapter 13 : Answers and Explanations

Answer Key

❋ Now, it's time to review your results to see where you went wrong and what areas you need to improve!

HiSET Mathematics Practice Tests

Practice Test 1

1	D	16	D	31	C	46	A
2	D	17	B	32	E	47	A
3	B	18	E	33	D	48	E
4	E	19	C	34	D	49	B
5	D	20	E	35	A	50	C
6	A	21	A	36	D		
7	C	22	D	37	C		
8	B	23	D	38	E		
9	B	24	C	39	D		
10	D	25	D	40	E		
11	A	26	C	41	D		
12	A	27	D	42	E		
13	C	28	E	43	B		
14	E	29	D	44	D		
15	A	30	A	45	C		

Practice Test 2

1	C	16	C	31	A	46	A
2	C	17	C	32	D	47	E
3	D	18	D	33	C	48	B
4	D	19	C	34	D	49	D
5	B	20	A	35	A	50	E
6	E	21	B	36	D		
7	E	22	C	37	C		
8	A	23	C	38	E		
9	D	24	E	39	B		
10	B	25	C	40	A		
11	D	26	D	41	C		
12	D	27	A	42	A		
13	B	28	B	43	C		
14	E	29	B	44	B		
15	A	30	E	45	D		

WWW.MathNotion.Com

HiSET Subject Test – Mathematics

HiSET Subject Test – Mathematics

Score Your Test

HiSET scores are reported on a 1–20 score scale in 1-point increments. Each subject test should be passed individually. It means that you must get 8 on each section of the test. If you failed one subject test but did well enough on another to get a total score of 45, that's still not a passing score.

You passed the test if you met the three HiSET passing criteria:

✓ Scored at least 8 out of 20 on each subtest.

✓ Scored at least 2 out of 6 on the essay.

✓ Achieved a total scaled score on all five HiSET subtests of at least 45 out of 100.

here are approximately 50 Multiple-choice questions on Mathematics section. Use the following table to convert HiSET Mathematics raw score to scaled score. (You get 1 point for each correct answer. No points for wrong or skipped answers.)

HiSET Mathematics raw score to scaled score	
Raw Scores	**Scaled Scores**
Below 22 (not passing)	Below 8
22 – 30	8 – 10
31 – 36	11 – 13
37 – 44	14 – 16
Above 44	*Aboe* 16

HiSET Subject Test – Mathematics

HiSET Subject Test – Mathematics

Practice Test 1
HiSET Mathematics

1) Answer: D

Let x be the number. Write the equation and solve for x.

$\frac{2}{5} \times 60 = \frac{4}{7} \cdot x \Rightarrow \frac{2 \times 60}{5} = \frac{4x}{7}$, use cross multiplication to solve for x.

$14 \times 60 = 4x \times 5 \Rightarrow 840 = 20x \Rightarrow x = 42$

2) Answer: D

If 10 balls are removed from the bag at random, there will be five ball in the bag.

The probability of choosing a white ball is 5 out of 25. Therefore, the probability of not choosing a white ball is 10 out of 25 and the probability of having not a white ball after removing 10 balls is the same.

3) Answer: B

$5^4 = 5 \times 5 \times 5 \times 5 = 625$

4) Answer: E

Write the numbers in order:

6, 12, 21, 21, 26, 39, 43, 50, 50

Since we have 9 numbers (9 is odd), then the median is the number in the middle, which is 26.

5) Answer: D

To find the discount, multiply the number by (100% – rate of discount).

Therefore, for the first discount we get: $(100\% - 20\%)(E) = (0.80)E$

For increase of 15 %:

$(0.80)E \times (100\% + 15\%) = (0.80)(1.15) = 0.92E$.

6) Answer: A

Use percent formula: $Part = \frac{percent \times whole}{100}$

$44 = \frac{percent \times 40}{100} \Rightarrow \frac{44}{1} = \frac{percent \times 40}{100}$, cross multiply.

$4,400 = percent \times 40$, divide both sides by 40: $110 = percent$

WWW.MathNotion.Com

HiSET Subject Test – Mathematics

7) Answer: C

The weight of 7.5 meters of this rope is: $7.5 \times 520g = 3,900g$

1 kg = 1,000 g, therefore, $3,900 \, g \div 1,000 = 3.9 kg$

8) Answer: B

$x = 42 + 108 = 150$

9) Answer: B

Some of prime numbers are: 2, 3, 5, 7, 11, 13

Find the product of two consecutive prime numbers:

$5 \times 7 = 35$ (not in the options)

$7 \times 11 = 77$ (not in the options)

$11 \times 13 = 143$ (not in the options)

$13 \times 17 = 221$ (bingo!)

Choice B is correct.

10) Answer: D

Use the formula for Percent of Change: $\frac{New\ Value - Old\ Value}{Old\ Value} \times 100\ \%$

$\frac{32-80}{80} \times 100\ \% = -60\ \%$

(negative sign here means that the new price is less than old price).

11) Answer: A

If the score of Harper was 24, therefore the score of Emma is 12. Since, the score of Zoe was one fourth of Emma, therefore, the score of Zoe is 3.

12) Answer: A

The area of the floor is: $4 \, cm \times 18 \, cm = 72 \, cm^2$

The number of tiles needed $= 72 \div 6 = 12$

13) Answer: C

Use FOIL (First, Out, In, Last)

$(x - 2y)(3x - 5y) = 3x^2 - 5xy - 6xy + 10y^2 = 3x^2 - 11xy + 10y^2$

14) Answer: E

The average speed of Ryan is: $140 \div 7 = 20 \, km$

HiSET Subject Test – Mathematics

The average speed of Riley is: $240 \div 8 = 30$ km

Write the ratio and simplify. $20:30 \Rightarrow 2:3$.

15) Answer: A

The sum of supplement angles is 180. Let x be that angle. Therefore,

$x + 5x = 180$.

$6x = 180$, divide both sides by 6: $x = 30$.

16) Answer: D

Use percent formula:

$Part = \frac{percent \times whole}{100}$

$286 = \frac{percent \times 440}{100} \Rightarrow$ (cross multiply): $28{,}600 = percent \times 440 \Rightarrow percent = \frac{28{,}600}{440} = 65$

286 is 65 % of 440. Therefore, the discount is: $100\% - 65\% = 35\%$.

17) Answer: B

$average = \frac{sum\ of\ terms}{number\ of\ terms} \Rightarrow average = \frac{(32+25+34+19)}{4} \Rightarrow \frac{110}{4} = 27.5$

18) Answer: E

Let x be the number. Write the equation and solve for x.

$\frac{(54-x)}{x} = 8$ (cross multiply)

$(54 - x) = 8x$, then add x both sides. $54 = 9x$, now divide both sides by 9. $\Rightarrow x = 6$.

19) Answer: C

Use Pythagorean Theorem: $a^2 + b^2 = c^2$

$15^2 + 20^2 = C^2 \Rightarrow 225 + 400 = C^2 \Rightarrow 625 = c^2 \Rightarrow c = 25$

20) Answer E

The area of the trapezoid is: $Area = \frac{1}{2}h(b_1 + b_2) = \frac{1}{2}(x)(25 + 30) = 330$

$\rightarrow 27.5x = 330 \rightarrow x = 12$

$y = \sqrt{5^2 + 12^2} = \sqrt{25 + 144} = \sqrt{169} = 13$

The perimeter of the trapezoid is: $12 + 25 + 13 + 30 = 80$

HiSET Subject Test – Mathematics

21) Answer: A

$15 \times 12 = \$180$

Petrol use: $5 \div 2 = 2.5, 12 \times 2.5 = 30$ liters

Petrol cost: $30 \times \$2.50 = \$75$

Money earned: $\$180 - \$75 = \$105$

22) Answer: D

Simplify:

$4(x - 2) + 36 = 3(x + 1)$

$4x - 8 + 36 = 3x + 3 \Rightarrow 4x + 28 = 3x + 3$

Subtract $3x$ from both sides:

$x + 28 = 3$, Subtract 28 to both sides: $x = 3 - 28 \Rightarrow x = -25$

23) Answer: D

If the length of the box is 72, then the width of the box is one ninth of it, 8, and the height of the box is 4 (half of the width). The volume of the box is:

$V = lwh = (72)(8)(4) = 2,304$

24) Answer: C

Let x be the original price.

If the price of the sofa is decreased by 20% to $592, then: 80% of $x = 592 \Rightarrow 0.80x = 592 \Rightarrow x = 592 \div 0.80 = 740$

25) Answer: D

The percent of girls playing tennis is: $70\% \times 40\% = 0.70 \times 0.40 = 0.28 = 28\%$

26) Answer: C

Use this formula: Percent of Change $= \frac{New\ Value - Old\ Value}{Old\ Value} \times 100\ \%$

$\frac{23,100 - 42,000}{42,000} \times 100\ \% = -45\ \%$ and $\frac{12,705 - 23,100}{23,100} \times 100\% = -45\ \%$

27) Answer: D

Use distributive property:

$4xy(x^2 + y^2) = 4xy(x^2) + 4xy(y^2) = 4yx^3 + 4xy^3$

WWW.MathNotion.Com

HiSET Subject Test – Mathematics

28) Answer: E

Let x be the smallest number. Then, these are the numbers:

$x, x+1, x+2, x+3, x+4$

Average $= \dfrac{\text{sum of terms}}{\text{number of terms}} \Rightarrow 42 = \dfrac{x+(x+1)+(x+2)+(x+3)+(x+4)}{5} \Rightarrow 42 = \dfrac{5x+10}{5}$

$\Rightarrow 42 = x + 2 \Rightarrow x = 40$

29) Answer: D

Area of the circle is less than 81π. Use the formula of areas of circles.

$Area = \pi r^2 \Rightarrow \pi r^2 < 81\pi \Rightarrow r^2 < 81 \Rightarrow r < 9$

Radius of the circle is less than 9. Let's put 9 for the radius. Now, use the circumference formula: Circumference $= 2\pi r = 2\pi (9) = 18\pi$

Since the radius of the circle is less than 9. Then, the circumference of the circle must be less than 18π. Choices D is less than 18π

30) Answer: A

Add the first 3 numbers. $72 + 63 + 45 = 180$

To find the distance traveled in the next 5 hours, multiply the average by number of hours. Distance $= Average \times Rate = 48 \times 5 = 240$

Add both numbers. $180 + 240 = 420$

31) Answer: C

Use simple interest formula: $I = prt$ (I= interest, p = principal, r = rate, t = time)

I = $(5,000)(0.0225)(8) = 900$

32) Answer: E

$y = 2c^3 - 8cd + 5d^2$

Plug in the values of a and b in the equation: $c = -2$ and $d = 2$

$y = 2(-2)^3 - 8(-2)(2) + 5(2)^2 = -16 - 8(-4) + 20 = -16 + 32 + 20 = 36$

33) Answer: D

The ratio of boy to girls is 4:3. Therefore, there are 4 boys out of 7 students. To find the answer, first divide the total number of students by 7, then multiply the result by 4.

$420 \div 7 = 60 \Rightarrow 60 \times 4 = 240$

HiSET Subject Test – Mathematics

34) Answer: D

The equation of a line is in the form of $y = mx + b$, where m is the slope of the line and b is the $y - intercept$ of the line.

Two points $(0, -1)$ and $(2, 7)$ are online A. Therefore, the slope of the line A is:

slope of line A $= \frac{y_2 - y_1}{x_2 - x_1} = \frac{7-(-1)}{2-0} = \frac{8}{2} = 4$

The slope of line A is 4. Thus, the formula of the line A is:

$y = mx + b = 4x + b$, choose a point and plug in the values of x and y in the equation to solve for b. Let's choose point $(2, 7)$. Then:

$$= 4x + b \rightarrow 7 = 8 + b \rightarrow b = 7 - 8 = -1$$

The equation of line A is: $y = 4x - 1$

Now, let's review the choices provided:

A. $(3, 9)$ $y = 4x - 1 \rightarrow 9 = 12 - 1 = 11$, This is not true.

B. $(1, 5)$ $y = 4x - 1 \rightarrow 5 = 4 - 1 = 3$, This is not true.

C. $(-1, -7)$ $y = 4x - 1 \rightarrow -7 = -4 - 1 = -5$ This is not true.

D. $(0, -1)$ $y = 4x - 1 \rightarrow -1 = 0 - 1 = -1$, This is true!

E. $(4, 12)$ $y = 4x - 1 \rightarrow 12 = 16 - 1 = 15$, This is not true.

35) Answer: A

Let x be the number of years. Therefore, $1,500 per year equals $1,500x$.

starting from $42,300 annual salary means you should add that amount to $1,500x$.

Income more than that is: $I > 1,500x + 42,300$

36) Answer: D

Rate of change $= \frac{y_2 - y_1}{x_2 - x_1}$

$f(x) = x^2 - 5$, from $x = 2$, to $x = 5$, then $f(2) = 2^2 - 5 = -1$, and $f(5) = 5^2 - 5 = 20$. $\frac{20-(-1)}{5-2} = \frac{21}{3} = 7$

37) Answer: C

average $= \frac{\text{sum of terms}}{\text{number of terms}}$

The sum of the high of all constructions is: $15 \times 140 = 2,100$ m

WWW.MathNotion.Com   184

HiSET Subject Test – Mathematics

The sum of the high of all towers is: $10 \times 160 = 1,600$ m

The sum of the high of all building is: $2,100 + 1,600 = 3,700$

$average = \frac{3,700}{25} = 148$

38) Answer: E

8% of the volume of the solution is alcohol. Let x be the volume of the solution. Then:

8% of x = 28 ml

$0.08\ x = 28 \Rightarrow \frac{8x}{100} = \frac{280}{10}$ cross multiply

$80x = 28,000 \Rightarrow$ (devide by 80) $x = 350$

39) Answer: D

Let x be the original price.

If the price of a laptop is decreased by 30% to $315, then:

$70\ \%\ of\ x = 315 \Rightarrow 0.70x = 315 \Rightarrow x = 315 \div 0.70 = 450$

40) Answer: E

Write the equation and solve for M:

0.90 F = 0.15 M, divide both sides by 0.15, then:

$\frac{0.90}{0.15}$ F = M, therefore: M = 6 F, and M is 6 times of F or it's 600% of F.

41) Answer: D

Frist factor the function:

$2x^3 - 4x^2 - 30x = 2x\ (x^2 - 2x - 15) = 2x(x - 5)(x + 3)$

To find the zeros, $f(x)$ should be zero. $f(x) = 2x(x - 5)(x + 3) = 0$

Therefore, the zeros are: $x = 0$

$(x - 5) = 0 \Rightarrow x = 5$

$(x + 3) = 0 \Rightarrow x = -3$

42) Answer: E

Use the information provided in the question to draw the shape.

Use Pythagorean Theorem: $a^2 + b^2 = c^2 \Rightarrow 120^2 + 160^2 = c^2$

$\Rightarrow 14,400 + 25,600 = c^2 \Rightarrow 40,000 = c^2 \Rightarrow c = 200.$

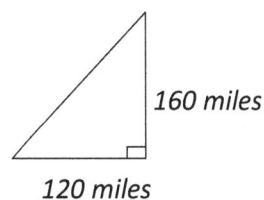

160 miles

120 miles

HiSET Subject Test – Mathematics

43) Answer: B

To find the number of possible outfit combinations, multiply number of options for each factor: $5 \times 2 \times 6 = 60$

44) Answer: D

Formula for the Surface area of a cylinder is:

$SA = 2\pi r^2 + 2\pi rh \rightarrow 72\pi = 2\pi r^2 + 2\pi r(5) \rightarrow r^2 + 5r - 36 = 0$

$(r+9)(r-4) = 0 \rightarrow r = 4$ or $r = -9$ (unacceptable)

45) Answer: C

To find the discount, multiply the number by (100% – rate of discount).

Therefore, for the first discount we get: $(900)(100\% - 22\%) = (900)(0.78)$

For the next 7% discount: $(900)(0.78)(0.93)$.

46) Answer: A

Use distance formula:

Distance $=$ Rate $\times$ time $\Rightarrow 228 = 40 \times T$, divide both sides by 40. $\Rightarrow T = 5.7$ hours. Change hours to minutes for the decimal part.

0.7 hours $= 0.7 \times 60 = 42$ $minutes$.

47) Answer: A

Let's compare each fraction:

$\frac{1}{4} = 0.25$; $\frac{5}{16} = 0.3125$; $\frac{4}{7} = 0.571$

$\frac{1}{4} < \frac{5}{16} < \frac{4}{7}$ Only choice A provides the right order.

48) Answer: E

$(4.2 \times 10^9) \times (3.5 \times 10^{-11}) = (4.2 \times 3.5) \times (10^9 \times 10^{-11}) = 14.7 \times (10^{(9-11)}) = 14.7 \times 10^{-2} = 0.147$

49) Answer: B

Solving Systems of Equations by Elimination

$\begin{array}{l} 2x + 3y = -8 \\ 3x - 4y = 5 \end{array}$ Multiply the first equation by 3, and second equation by -2, then add two equations.

HiSET Subject Test – Mathematics

$$\begin{matrix} 3(2x+3y=-8) \\ -2(3x-4y=5) \end{matrix} \Rightarrow \begin{matrix} 6x+9y=-24 \\ -6x+8y=-10 \end{matrix} \Rightarrow 17y=-34 \Rightarrow y=-2.$$

50) Answer: C

The formula of the volume of pyramid is:

$$V = \frac{l \times w \times h}{3}$$

The length and width of the pyramid is 5 cm, and its height is 15 cm. Therefore:

$$V = \frac{5 \times 5 \times 15}{3} = 125 \; cm^3$$

HiSET Subject Test – Mathematics

HiSET Subject Test – Mathematics

Practice Test 2
HiSET Mathematics

1) Answer: C

average (mean) = $\frac{\text{sum of terms}}{\text{number of terms}}$ ⇒ $65 = \frac{\text{sum of terms}}{90}$ ⇒ sum = $65 \times 90 = 5,850$

The difference of 73 and 37 is 36. Therefore, 36 should be subtracted from the sum.

$5,850 - 36 = 5,814$

mean = $\frac{\text{sum of terms}}{\text{number of terms}}$ ⇒ mean = $\frac{5,814}{90} = 64.6$

2) Answer: C

Solve for x. $-2 \leq 3x - 5 < 19$

⇒ (add 5 all sides) $-2 + 5 \leq 3x - 5 + 5 < 19 + 5$ ⇒ $3 \leq 3x < 24$

⇒ (divide all sides by 3) $1 \leq x < 8$

x is between 1 and 8. Choice C represent this inequality.

3) Answer: D

To get a sum of 5 for two dice, we can get 4 different options: $(1, 4), (4, 1), (2, 3), (3, 2)$

To get a sum of 8 for two dice, we can get 5 different options:

$(2, 6), (6, 2), (3, 5), (5, 3), (4, 4)$

Therefore, there are 9 options to get the sum of 5 or 8.

Since, we have $6 \times 6 = 36$ total options, the probability of getting a sum of 5 or 8 is 9 out of 36 or $\frac{9}{36} = \frac{1}{4}$.

4) Answer: D

Simplify and combine like terms.

$(6x^3 - 3x^2 - 5x^4) - (4x^2 - 2x^4 - 5x^3)$ ⇒ $(6x^3 - 3x^2 - 5x^4) - 4x^2 + 2x^4 + 5x^3$

⇒ $-3x^4 + 11x^3 - 7x^2 = -(3x^4 - 11x^3 + 7x^2)$.

5) Answer: B

The area of the square is 812.25. Therefore, the side of the square is square root of the area. $\sqrt{812.25} = 28.5$

Four times the side of the square is the perimeter: $4 \times 28.5 = 114$

HiSET Subject Test – Mathematics

6) Answer: E

Write the equation and solve for B: 0.60 A = 0.12 B, divide both sides by 0.12, then:

$\frac{0.60}{0.12}$ A = B, therefore: B = 5 A, and B is 5 times of A or it's 500% of A.

7) Answer: E

$f(x) = 2x^3 + 3x^2 - 2x + 9$, and $g(x) = 3$, then

$(f \circ g)(x) = f(g(x)) = f(3) = 2(3)^3 + 3(3)^2 - 2(3) + 9 = 54 + 27 - 6 + 9 = 84$

8) Answer: A

3,500 out of 56,000 equals to $\frac{3,500}{56,000} = \frac{35}{560} = \frac{5}{80} = \frac{1}{16}$

9) Answer: D

Write the numbers in order: 11, 16, 24, 32, 35, 41, 58

Median is the number in the middle. So, the median is 32.

10) Answer: B

Use simple interest formula: $I = prt$ (I = interest, p = principal, r = rate, t = time)

$I = (23,000)(0.0225)(6) = 3,105$

11) Answer: D

To find the zeros, $f(x)$ should be zero.

$f(x) = ax^2 + bx + c = 0 \Rightarrow x = \frac{-b \pm \sqrt{b^2 - 4ac}}{2a}$

$f(x) = 2x^2 + 4x - 48 = 0, \Rightarrow x = \frac{-4 \pm \sqrt{4^2 - 4(2)(-48)}}{2(2)} = \frac{-4 \pm \sqrt{400}}{4}$

$\Rightarrow x = \begin{cases} \frac{-4+20}{4} = \frac{16}{4} = 4 \\ \frac{-4-20}{4} = \frac{-24}{4} = -6 \end{cases}$

12) Answer: D

Six times of 15,000 is 90,000. One fifth of them cancelled their tickets.

One fifth of 90,000 equal 18,000 ($\frac{1}{5} \times 90,000 = 18,000$).

$(90,000 - 18,000 = 72,000)$ fans are attending this week

WWW.MathNotion.Com

HiSET Subject Test – Mathematics

13) Answer: B

the population is increased by 25% and 15%. 25% increase changes the population to 125% of original population.

For the second increase, multiply the result by 115%.

$(1.25) \times (1.15) = 1.438 = 143.8\%$

43.8 percent of the population is increased after two years.

14) Answer: E

Change the numbers to decimal and then compare.

$\frac{1}{12} = 0.083\ldots$

0.05

$7\% = 0.07$

$\frac{1}{25} = 0.04$

Therefore $\frac{1}{12} > 7\% > 0.05 > \frac{1}{25}$.

15) Answer: A

Volume of a box = length × width × height = $9 \times 6 \times 5 = 270$

16) Answer: C

The diagonal of the square is 8. Let x be the side.

Use Pythagorean Theorem: $a^2 + b^2 = c^2$

$x^2 + x^2 = 8^2 \Rightarrow 2x^2 = 64 \Rightarrow x^2 = 32 \Rightarrow x = \sqrt{32}$

The area of the square is: $\sqrt{32} \times \sqrt{32} = 32$

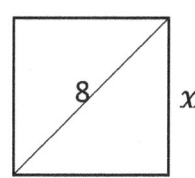

17) Answer: C

average = $\frac{\text{sum of terms}}{\text{number of terms}} \Rightarrow 25 = \frac{\text{sum of 8 numbers}}{8} \Rightarrow$ sum of 8 numbers = $25 \times 8 = 200$

$29 = \frac{\text{sum of 6 numbers}}{6} \Rightarrow$ sum of 6 numbers = $6 \times 29 = 174$

sum of 8 numbers − sum of 6 numbers = sum of 2 numbers

$200 − 174 = 26$

average of 2 numbers = $\frac{26}{2} = 13$.

HiSET Subject Test – Mathematics

18) Answer: D

Solving Systems of Equations by Elimination

$4x + 5y = -12$
$4x + 2y = 6$ Multiply the first equation by -1,

then add two equations.

$-1(4x + 5y = -12)$ $\Rightarrow$ $-4x - 5y = 12$ $\Rightarrow$ $-3y = 18 \Rightarrow y = -6$.
$(4x + 2y = 6)$ $\quad\quad 4x + 2y = 6$

19) Answer: C

Write a proportion and solve for x.

$\frac{9}{11} = \frac{63}{x} \Rightarrow 9x = 11 \times 63 \Rightarrow x = 77$ in

20) Answer: A

The width of the rectangle is triple its length. Let x be the length. Then, $width = 3x$

Perimeter of the rectangle is 2 (width + length) = $2(3x + x) = 56 \Rightarrow 8x = 56 \Rightarrow x = 7$

Length of the rectangle is 7 meters.

21) Answer: B

In the stadium the ratio of home fans to visiting fans in a crowd is 9:4. Therefore, total number of fans must be divisible by 13: $9 + 4 = 13$.

Let's review the choices:

A. 52,630:   52,630 ÷ 13 = 4,048.46

B. 41,249:   41,249 ÷ 13 = 3,173

C. 34,620:   34,620 ÷ 13 = 2,663.08

D. 72,936:   72,936 ÷ 13 = 5,610.46

E. 48,123:   48,123 ÷ 13 = 3,701.77

Only choice B when divided by 13 result a whole number.

22) Answer: C

Plug in 149 for F and then solve for C.

$C = \frac{5}{9}(F - 32) \Rightarrow C = C = \frac{5}{9}(149 - 32) \Rightarrow C = \frac{5}{9}(117) = 65$.

HiSET Subject Test – Mathematics

23) Answer: C

$5x - 2y = 1$. Plug in the values of x and y from choices provided. Then:

A. $(2, 1)$: $5(2) - 2(1) = 10 - 2 = 8$, This is NOT true!

B. $(-1, 3)$: $5(-1) - 2(3) = -5 - 6 = -11$, This is NOT true!

C. $(1, 2)$: $5(1) - 2(2) = 5 - 4 = 1$, This is true!

D. $(-2, 2)$: $5(-2) - 2(2) = -10 - 4 = -14$, This is NOT true!

E. $(3, 0)$: $5(3) - 2(0) = 15 - 0 = 15$, This is NOT true!

24) Answer: E

To find the number of possible outfit combinations, multiply number of options for each factor: $5 \times 9 \times 4 = 180$

25) Answer: C

Use FOIL method.

$(x - 3y)^2 = (x - 3y)(x - 3y) = x^2 - 3xy - 3xy + 9y^2 = x^2 - 6xy + 9y^2$

26) Answer: D

$$\text{Probability} = \frac{\text{number of desired outcomes}}{\text{number of total outcomes}} = \frac{20}{28+21+20+11} = \frac{20}{80} = \frac{1}{4}$$

27) Answer: A

First, find the sum of six numbers.

$\text{average} = \frac{\text{sum of terms}}{\text{number of terms}} \Rightarrow 70 = \frac{\text{sum of 6 numbers}}{6} \Rightarrow \text{sum of 6 numbers} = 6 \times 70 = 420$

The sum of 6 numbers is 420. If a seventh number that is greater than 77 is added to these numbers, then the sum of 7 numbers must be greater than 497 ($420 + 77 = 497$). If the number was 77, then the average of the numbers is:

$\text{average} = \frac{\text{sum of terms}}{\text{number of terms}} = \frac{497}{7} = 71$

Since the number is bigger than 77. Then, the average of seven numbers must be greater than 71. Choices A is greater than 71.

28) Answer: B

The perimeter of the trapezoid is 52 cm.

Therefore, the missing side (height) is $= 52 - 10 - 16 - 12 = 14$

HiSET Subject Test – Mathematics

Area of a trapezoid: A = $\frac{1}{2}$ h (b1 + b2) = $\frac{1}{2}$ (14) (10 + 12) = 154

29) Answer: B

The probability of choosing a spade or hearts is $\frac{26}{52} = \frac{1}{2}$

30) Answer: E

Plug in the value of x and y. $-5(2x + y) + (6 - x)^2$ when $x = 2$ and $y = -3$

$= -5(2(2) + (-3)) + (6 - (2))^2 = -5(4 - 3) + (6 - 2)^2 = (-5)(1) + (4)^2 =$
$-5 + 16 = 11$

31) Answer: A

The ratio of boy to girls is 3:7. Therefore, there are 3 boys out of 10 students. To find the answer, first divide the total number of students by 10, then multiply the result by 3.
$120 \div 10 = 12 \Rightarrow 12 \times 3 = 36$

There are 36 boys and 84 (120 – 36) girls. So, 48 more boys should be enrolled to make the ratio 1:1

32) Answer: D

Surface Area of a cylinder = $2\pi r$ (r + h),

The radius of the cylinder is 4 (8 ÷ 2) inches and its height is 16 inches. Therefore,
Surface Area of a cylinder = 2π (4) (4 + 16) = 160π

33) Answer: C

Simplify. $3x^3y^4(-2xy^2)^2 = 3x^3y^4(4x^2y^4) = 12x^5y^8$

34) Answer: D

The distance between Daniel and Noa is 54 miles. Daniel running at 4.5 miles per hour and Noa is running at the speed of 9 miles per hour. Therefore, every hour the distance is 4.5 miles less. $54 \div 4.5 = 12$.

35) Answer: A

Let x be the number of new shoes the team can purchase. Therefore, the team can purchase 160 x.

The team had $56,000 and spent $36,000. Now the team can spend on new shoes $20,000 at most. Now, write the inequality: $160x + 36,000 \leq 56,000$

HiSET Subject Test – Mathematics

36) Answer: D

The failing rate is 24 out of 75, $\frac{24}{75}$

Change the fraction to percent: $\frac{24}{75} \times 100\% = 32\%$

32 percent of students failed. Therefore, 68 percent of students passed the exam.

37) Answer: C

Let L be the length of the rectangular and W be the width of the rectangular. Then,

$L = 7W - 4$

The perimeter of the rectangle is 96 meters. Therefore: $2L + 2W = 88$

$L + W = 44$

Replace the value of L from the first equation into the second equation and solve for W:

$(7W - 4) + W = 44 \to 8W - 4 = 44 \to 8W = 48 \to W = 6$

The width of the rectangle is 4 meters, and its length is:

$L = 7W - 4 = 7(6) - 4 = 38$

The area of the rectangle is: length × width = 38 × 6 = 228

38) Answer: E

Let x be the number. Write the equation and solve for x.

80% of $x = 36 \Rightarrow 0.80\, x = 36 \Rightarrow x = 36 \div 0.80 = 45$

39) Answer: B

$\frac{80}{125}$, simplify by 5, then the number is the square root of $\frac{16}{25}$: $\sqrt{\frac{16}{25}} = \frac{4}{5}$

The cube of the number is: $(\frac{4}{5})^3 = \frac{64}{125}$

40) Answer: A

Let x be all expenses, then $\frac{20}{100} x = \$640 \to x = \frac{100 \times \$640}{20} = \$3,200$

He spent for his rent: $\frac{30}{100} \times \$3,200 = \$960$

41) Answer: C

Let x be the capacity of one tank. Then, $\frac{3}{5}x = 420 \to x = \frac{5 \times 420}{3} = 700\ L$

The amount of water in three tanks is equal to: $3 \times 700 = 2,100$ Liters

HiSET Subject Test – Mathematics

42) Answer: A

To solve absolute values equations, write two equations.

$-4x + 8$ can equal positive 36, or negative 36. Therefore,

$-4x + 8 = 36 \Rightarrow -4x = 28 \Rightarrow x = -7$

$-4x + 8 = -36 \Rightarrow -4x = -44 \Rightarrow x = 11$

43) Answer: C

First, find the number.

Let x be the number. Write the equation and solve for x.

140% of a number is 126, then:

$1.4 \times x = 126 \Rightarrow x = 126 \div 1.4 = 90$

40% of 90 is: $0.4 \times 90 = 36$

44) Answer: B

Use formula of rectangle prism volume.

V = (length) (width) (height) $\Rightarrow$ 6,300 = (30) (15) (height) $\Rightarrow$ height = 6,300 ÷ 450 = 14

45) Answer: D

Isolate and solve for x.

$\frac{5}{12}x + \frac{1}{6} = \frac{3}{4} \Rightarrow \frac{5}{12}x = \frac{3}{4} - \frac{1}{6} = \frac{7}{12} \Rightarrow \frac{5}{12}x = \frac{7}{12}$

Multiply both sides by the reciprocal of the coefficient of x: $(\frac{12}{5})\frac{5}{12}x = \frac{7}{12}(\frac{12}{5}) \Rightarrow x = \frac{7}{5}$

46) Answer: A

Mrs. Thomson needs an 76% average to pass for four exams. Therefore, the sum of 5 exams must be at lease $5 \times 76 = 380$

The sum of 4 exams is: $78 + 81 + 82 + 89 = 330$

The minimum score Mrs. Thomson can earn on her fifth and final test to pass is: $380 - 330 = 50$

47) Answer: E

Use Pythagorean Theorem: $a^2 + b^2 = c^2$

$18^2 + 24^2 = c^2 \Rightarrow 900 = c^2 \Rightarrow c = 30$

HiSET Subject Test – Mathematics

48) Answer: B

The equation of a line in slope intercept form is: $y = mx + b$

Solve for y.

$7x - 2y = 8 \rightarrow -2y = -7x + 8$

Divide both sides by (-2). Then: $y = \frac{7}{2}x + 4$

The slope of this line is $\frac{7}{2}$.

The product of the slopes of two perpendicular lines is -1. Therefore, the slope of a line that is perpendicular to this line is:

$m_1 \times m_2 = -1 \Rightarrow \frac{7}{2} \times m_2 = -1 \Rightarrow m_2 = \frac{-1}{\frac{7}{2}} = -\frac{2}{7}$

49) Answer: D

$(8,4), (5,3), (2,2), (4,1)$

A function is a relation in which each element in the domain corresponds to exactly one element of the range. Then, option D is function.

50) Answer: E

Use PEMDAS (order of operation):

$49 + 7 \times (-16) \div 14 - 9 = 49 + (-112 \div 14) - 9 = 49 + (-8) - 9 = 32.$

"End"

www.ingramcontent.com/pod-product-compliance
Lightning Source LLC
Chambersburg PA
CBHW080438110426
42743CB00016B/3202